Death on the Job

Daniel M. Berman

| Death on the Job |

*Occupational Health
and Safety Struggles
in the United States*

Monthly Review Press
New York and London

Library of Congress Cataloging in Publication Data
Berman, Daniel M.
  Death on the job.
  Includes bibliographical references and index.
  1. Industrial hygiene—United States. 2. Industrial safety—United
States. 3. Trade unions—United States. 4. Workmen's compensation—
United States. 5. Work environment—United States. I. Title. [DNLM:
1. Occupational health services—History—United States. 2. Occupa-
tional health services—United States—Legislation. 3. Accidents,
Industrial—Prevention and control—United States—Legislation.
4. Industrial medicine—History—United States. 5. Industrial
medicine—United States—Legislation. WA11 AA1 B47d]
HD7654.B47   614.8′52′0973   78-13914
ISBN 0-85345-462-0

Manufactured in the United States of America
10 9 8 7 6 5 4

To my family and mother, Dr. Jane Montzingo Berman,
my first teacher about politics and health,
and to Jake Jaundzems (1941–1978)

Contents

Preface
and Acknowledgments

Business has been forced to deal with issues of occupational safety and health under two different sets of conditions: during war-imposed labor shortages and during periods of severe social upheaval. When labor is extremely scarce, employers worry about preserving the labor they control by making work more attractive. During times of severe social unrest workers demand better conditions. In both cases business tries to jump ahead of the workers and create institutions which define the problems of health and safety in nonthreatening ways and take the sting out of the workers' unrest. Perhaps it should not be surprising that the new occupational health and safety movement arose during an epoch that combined labor shortages and widespread protest against the unpopular war in Vietnam.

In the early twentieth century U.S. corporations responded to concern about work accidents by setting up a business-controlled compensation-safety apparatus which held down compensation costs and did little to improve working conditions. This apparatus was able to exclude the issue of occupational health and safety from open debate until the late 1960s through its control of research, education, workers' compensation, and government appointments, and by creating the public impression that health problems in the workplace were almost nonexistent. As a result, the pain and bloodshed and nearly all the money costs of work-related diseases and injuries are still borne by workers and their families and the public at large—that way it's cheaper for industry.

With the exceptions of the United Mine Workers' activities and sporadic local uprisings, unions have been seriously involved in health and safety only in the last decade, since they mobilized to pass the Occupational Safety and Health Act of 1970. The OSHA law was made possible by a tight labor market, worker dissatisfaction, the new environmental consciousness, the aid of progressive professionals, and a climate of general social unrest. Although business interests, through a complex compensation-safety apparatus, have been able to dominate the operation of the new governmental institutions created to regulate working conditions, the question of occupational health and safety is now on the permanent agenda of workers, their unions, and the public.

In the tradition of Crystal Eastman's *Work-Accidents and the Law* (1910) I have tried to write a technical work which could be read by workers, students, interested professionals, and the public at large. The controversial nature of some of the conclusions has led me to document every argument in meticulous detail, but the general reader can skip the tables in Appendix 1 and the footnotes with no loss in understanding. As an organizer and pamphleteer in the field I know the necessity of getting the facts straight, and I hope what is written here will furnish tools to save lives and widen the struggle over working conditions. Appendix 2 presents a "Short Guide to Worker-Oriented Sources in Occupational Safety and Health" in an effort to devise counterinstitutions which will function over the long term on behalf of working people rather than of the corporations.

In 1975-1976 I was the fortunate beneficiary of a Milbank postdoctoral fellowship at Boston University. Since then I have supported myself by jobs with the Montana Energy and MHD Research and Development Institute, the Massachusetts Division of Employment Security, and the Internationales Institut für Vergleichende Gesellschaftsforschung in Berlin. I have also bottled Lucky beer and Dr. Pepper, driven a truck, and worked as a laborer with Dragger's Gulch Temporary Restorations in San Francisco.

It would have been impossible to write this book without the help of scores of people in the United States and abroad who believed in its importance and kept me informed about the latest events. Stu Leiderman got me started on environmental issues in 1969, and Mike Ryan and Art Button of Teamsters Local 688 provided me with my first chance to investigate and organize around working conditions in St. Louis. Professors Jim Davis and Dennis Judd enthusiastically criticized an early draft of this manuscript, which earned me a doctorate at Washington University. Quentin Young, Phyllis Cullen, and Don Whorton saw to it that I was hired as director of the Occupational Health Project of the Medical Committee for Human Rights, which put me in the center of the movement in 1972, right after the formation of the Chicago Area Committee for Occupational Safety and Health (CACOSH). John McKinlay arranged time for me to start writing this book at Boston University. I would like to thank Carl Carlson, Will Shortell, Dave Simmons, Rick Engler, Jim Weeks, Dick Ginnold, and Sharon Itaya for keeping me up to date in their respective cities; and Dick Marco and Bob Fowler for their unique insights on raising the issues of working conditions in local unions. Frank Wallick, Tony Mazzocchi, Steve Wodka, and Bert Cottine shepherded me around Washington before I could find my own way, and Dave Kotelchuk and Frank Goldsmith did the same for me in New York.

Irving J. Selikoff and Philip S. Foner, giants in their fields, read the early chapters and helped me avoid some errors of fact. Andrew Kalmykow and Alan Tebb graciously helped me track down information about the insurance industry which would have been hard to come by without their help. David Bacon diagramed the entire manuscript for me at an early date, and Andrea Hricko, Morris Davis, Peter Goose, Jeannette Harris, Scott McAllister, Gordie Gunderson, Patricia Cesta, and Phil Polakoff answered questions nobody else could. Barry Castleman, Jekabs Jaundzems, Steve Blumfield,

Charlie Clutterbuck, Pedro Comedor de Abroba, Vicente Navarro, Rodney Larson, Frieder Naschold, Smaro Chiotras, Madaline Jaundzems, and Jettie Harrison forced me to think hard about the international situation; and my talks with Molly Coye, Josephine Ann Firth, Roberto Dutra, Eduardo Limpo, Kathleen Ramos, and Ellen Widess helped clarify the final chapter in my mind. I owe special thanks to Karen Judd of Monthly Review Press for her painstaking editing, especially of the last three chapters, and to Judy Ruben for her faith that the book would finally appear. Finally, I want to show my gratitude to those who helped me understand the problems of working conditions and who fear being mentioned by name. Perhaps this book will help cut through a bit of the fear. Naturally, none of the people mentioned above bears any responsibility for the final form of the book.

Daniel M. Berman
September 16, 1978

|1|

Why Work Kills

Killing Him Softly—The Story of Marcos Vela

Marcos Vela began working as a machine tender for the Johns-Manville asbestos factory in Pittsburg, California around 1935. In 1959 the company started a policy of periodic medical examinations concentrating on the detection of lung disease. That year a private physician paid by Johns-Manville obtained a chest x-ray on Vela and noted a finding of occupationally related disease. The report contained no recommendation about changing the work environment, and Vela was not informed that he was developing asbestosis.

In 1962 Vela was examined by Dr. Kent D. Wise, another company-paid physician. A chest film taken at his office showed the presence of lung disease. As before, the patient was told nothing. Dr. Wise saw Marcos Vela again in 1965 for another routine examination and sent him to a nearby hospital for another chest x-ray. The radiologist made a diagnosis of work-related pneumoconiosis, which he communicated to Dr. Wise. Vela was told nothing.

In February 1968, at his fourth routine physical for Johns-Manville, Marcos Vela complained of coughing and shortness of breath. Though his x-ray showed a "ground glass appearance," Vela was told by the company nurse that "everything was fine." Again he was not informed of the adverse medical findings by Dr. Wise. That August he was hospitalized, unable to catch his breath. He would never return to work.

Thus for ten years doctors under contract to the Johns-Manville Corporation, the world's largest private producer of asbestos, knew that one of their workers was developing asbestosis, and not only refused to tell him, but took no measures to prevent further exposure to the asbestos which was ruining his health. That same company had known since 1931, from research it had sponsored, that asbestos caused lung disease.

*On the suggestion of his workers' compensation attorney, Marcos Vela sued Dr. Wise for medical malpractice, and personally insisted on taking the case to court. Expert witnesses from both sides testified that the Vela x-rays showed evidence of developing lung disease and that it had been the physician's duty to inform the patient of that fact and to take effective action to secure treatment and prevent the disease from developing further. In November 1973, a Contra Costa County jury agreed with the worker-plaintiff that Dr. Wise had committed medical malpractice, and awarded Marcos Vela damages of $351,000 against the doctor. On January 2, 1975, Vela received a check for $208,506.30, and his attorneys collected another $170,596.06 in fees from Dr. Wise's insurance company.**

Marcos Vela is a slight, courtly man who has to move around and gesture in slow motion. His parents brought him from Mexico when he was very young, and he still likes to tell jokes in Spanish, though married to an anglo-sajona. *One woman who met him said he must have looked like Rudolph Valentino when he was younger and that he still had the smile and the eyes.*

He took the job with Johns-Manville when steady jobs were hard to find, especially for Mexican-Americans, and worked in clouds of asbestos dust for decades.

Sometimes you couldn't see across the shop. Back in 1958, 1960, if you asked for a mask they say it's a crime. The super-

*Quantities add up to more than $351,000 because of accrued interest.

intendent told me: 'The only thing about this dust is that it's uncomfortable; it hasn't ever hurt anybody.' He's dead now; he mighta had it. Many of them are gone; there's two or three that are pretty sick. My own brother's got it real bad. They shut down my old department about two years ago.

In May 1968 when I was about to quit, [the company] got a group of us together and said we had to wear those masks, but they didn't say why. Then later Dr. Wise said that if you wear this mask for four or five years it will clear up the stuff in your lungs. When I had to quit the company did nothing for me. They said, 'Go look for another job.' And the union . . . somehow the guys didn't seem to be interested.

Marcos Vela now wears a $100 gold watch inscribed:

Marcos A. Vela, presented by Johns-Manville Corporation in appreciation of 25 years of loyal service, October 1960.

His lung capacity is down to one-fourth of normal and he stops to take rapid shallow breaths when he speaks quickly. He consumes eighteen kinds of medicine daily, and has a respirator connected to an oxygen tank in the bedroom. Some of the pills erode the enamel of his teeth, so he takes other pills to counteract that effect.

The doctors didn't expect me to last this long. I can't run; I have to walk slow, though I'm not crippled completely. I have a bulb of compressed oxygen with me, in case of coughing spells. As soon as I have a cold I have to go to the hospital. Once I had a fever of half a degree and had to stay in the hospital for three weeks. The doctor says I have to be careful about shaking hands with people, because of the germs, and I can't stand smokers or smoggy days, when I have to stay inside with the air filter on. It's hard on my family; with all this money I have I can't get my health back.[1]

How could such an atrocity occur? What system made it possible for a physician to neglect to treat a progressive, disabling, and potentially fatal illness; to refuse to tell the sick man about it; and to avoid taking any steps which would prevent the progression of the disease? Why did a company *which knew about the dangers of asbestos for at least three decades allow exposure to lethal levels of asbestos dust?*

Where were the insurance company and state and federal inspectors who were supposed to ensure healthful conditions on the job? Why did the union allow its members to work under such hazardous conditions? Is Marcos Vela's story being repeated in workplaces throughout the United States, or is his case a rare anomaly caused by wicked managers and a heartless physician? Must the everyday functioning of American capitalism destroy Marcos Velas as inevitably as it produces Chevrolets?

Introduction

In the first two decades of this century, monopoly corporations* such as U.S. Steel responded to the movement around occupational safety and health by setting up a business-controlled "compensation-safety apparatus," a stalling operation which, by appearing to be doing something, withheld the issue of working conditions from the public agenda until the late 1960s. The privately owned workers' compensation system founded between 1910 and 1920 in most states gave minimal compensation to many workers disabled by work accidents. The skeleton implementation of workers' compensation and state-run industrial inspection programs was enough to create the public impression that working conditions were being improved as quickly as possible and that injured workers were being helped satisfactorily (see chapter 2).

In fact, the workers' compensation system stabilized compensation costs to employers at 1 percent of payroll by

---

*Monopoly sector corporations largely control the prices of their products through control of most of the production, either singly or in cooperation with other giants in an industry. Typically, they produce for a national or international market, are capital-intensive, and since the 1930s (in the United States) are unionized.[2]

almost totally ignoring the problems of long-term disability, occupational disease, and worker rehabilitation. By setting up a closed compensation bureaucracy, companies avoided the costs and embarrassments of jury trials. Chapter 3 shows how the compensation system, in effect, shifts at least 90 percent of the financial burdens of work-related casualties onto workers, their families, and the public, while government and private agencies ostensibly designed to prevent injuries largely ignore such tasks.

Chapter 4 documents the fact that, to minimize financial settlements to workers, companies created the ghettoized institution of company medicine, in which the industrial physician became the company's advocate in compensation claims, backed by an infrastructure of lawyers and corporate-sponsored research findings that discounted job hazards.

Only in the last ten years has independent research, often with union support, begun systematically questioning the doctrines and practices of the "compensation-safety apparatus." Despite the occurrence of sporadic rank-and-file uprisings over working conditions, it has only been since the late 1960s, a time of new environmental consciousness and general social unrest, that unions have been strongly involved in the questions of job health and safety. Concern over health and safety has helped provoke radical reform within a few unions. The political movement that created the Occupational Safety and Health Act of 1970 (OSHA)* has called increasing attention to the importance of preventing casualties. Chapter 5 explores efforts by activists from unions and the professions to look into apparatus efforts to blame "accident-prone" workers, and, instead, to concentrate on speedup, sloppy maintenance, worker powerlessness, and faulty industrial design as causes of work injury and disease.

Although the corporate elite, through the compensation-safety apparatus, has been able to dominate the operation of the federal institutions created by the 1970 law, working conditions are now on the permanent agenda of workers,

---

*The acronym OSHA refers both to the law itself and to the Occupational Safety and Health Administration created to enforce it.

unions, and the U.S. public. Chapter 6 examines recent developments and outlines prescriptions for the future.

The Political-Economic Context of Industrial Safety at the Turn of the Century

The rapid industrialization of the United States produced a multitude of new dangers for workers. Big business, unable to control "ruinous" competition and confronting a militant working class and a growing socialist movement, sought the aid of the federal government. The fruits of fabulous productivity increases were gradually concentrated in fewer and fewer hands, symbolized in 1901 by the organization of the United States Steel Corporation, the nation's first billion-dollar business. As a result, the competitive sector, i.e., independent farmers and small businesses, was squeezed by the relentless advance of big business allied to the banks. Smaller manufacturers, unable to raise prices easily, violently fought unionization, while a few leaders from the monopoly sector began to devise sophisticated methods to forestall unionization through token "welfare" policies.[3]

Farming became more expensive as more machines were used per worker, and small farmers, hopelessly in debt, often lost their land.[4] In the South, black people were pushed to the edge of survival by the reimposition of legal segregation, loss of the vote, and systematic expropriation of what little property they had by the combined use of sharecropping and the lynching, jailing, and expropriation of "uppity" blacks.[5] Indians who had survived the "winning of the West" were locked in reservations, and U.S. troops were at the same time fighting Filipinos in the Philippines in the drive to open new areas to capitalist exploitation.

Meanwhile, by the end of the nineteenth century, large corporations and sweatshops began employing millions of hopeful immigrants in dirty jobs that still paid better than peasant work in Southern and Eastern Europe. Giant corporations, led by the railroads, usually learned to use regulatory commissions to consolidate their control of markets and public opinion. In 1893, six years after the formation of the Interstate Commerce Commission, former U.S. Attorney-General Richard Olney wrote:

> The Commission ... can be ... of great use to the railroads. It satisfies the popular clamor for government supervision of the railroads at the same time that that supervision is almost entirely nominal. ... It thus becomes a ... barrier between the railroads and the people and a sort of protection against hasty and crude legislation hostile to railroad interests.[6]

The Congress and presidency were finally secured for big business by the election in 1896 of William McKinley over the ragtag Democratic-Populist coalition.[7]

Labor, badly beaten in a series of strikes in the late 1880s and early 1890s, was on the ascendancy by 1900, but unions made little progress in organizing the new mass-production industries. Unions had been practically eliminated in the steel industry by the turn of the century. The coal and hard-rock mines, the brewing industry, and the garment trades, where industrial unions* had managed to make permanent inroads, were scenes of bloody struggles. Between 1890 and 1914, the few unionized craft workers and the unorganized majority saw their wage increases negated by inflation, while many of their efforts to organize were stymied by the open-shop campaign of the National Association of Manufacturers, representing smaller manufacturers in the competitive sector.[8] It is no wonder that the Industrial Workers of the World (the famous "Wobblies") won a couple of spectacular strikes, that strenuous attempts to organize unions resulted in the enlistment of thousands of workers, and that millions of voters turned to the Socialist Party.[9]

Work in those days was hard and dangerous for most; hours were extremely long, and nonunionized workers were

---

*Industrial unions organize all workers in a workplace into one union, rather than dividing them according to craft. The Carpenters Union is a craft union; the United Auto Workers Union is an industrial one.

largely unable to fight back against employers' attempts to extract more production.[10] A 1904 report in the labor press estimated that 27,000 workers were killed on the job each year,[11] and a 1907 Bureau of Labor report put the annual death toll at 15,000 to 17,500 of 26 million male workers.[12] Although the figures are difficult to interpret, it is likely that the on-the-job death rate at the turn of the century was at least double the present rate. An idea of just how dangerous work was is reflected in such contemporary accounts as that by Mack Sennett, the inventor of the Keystone Kops, who started working at age seventeen for the American Iron Works in East Berlin, Connecticut.

> In my day it was common for four men to hoist a four-hundred-pound rail, place it on the shoulders of a single man, and expect him to tote it a hundred yards.
>
> I . . . loathed a "bucker-upper" named Smith, who caught white-hot rivets in a bucket as fast as he could—he was a pieceworker . . . and ordered me to slam them home at high speed with a ten-pound sledge hammer. The faster I slammed, the more money Mr. Smith made.
>
> An ironworks is a hot place. Iron is produced from ores, hematite, magnetite, limestone, and siderite. This is heated in smelters, some of them one hundred feet tall, and poured in a molten liquid often as hot as 3500 degrees. When you ladle the slag of a pool of iron it sometimes sputters like fat on the stove. My hands are scarred with white marks from doing that when I was seventeen.[13]

Although labor made tremendous efforts for shorter hours, through strikes and legislation, the attempts usually ended in failure. Work weeks surpassed belief. In 1895 the New York State commissioner of labor recommended a law to reduce bakers' hours. His official report stated:

> Were it not fully established by indisputable proof, credence would hardly be given to the fact that many bakers are obliged to work more than one hundred hours per week, and in some instances they labor one hundred and thirty-two hours in a week. These unnecessary hours . . . as the investigation demonstrates,

have resulted in many cases in great physical injury to the employees.[14]

Later, the U.S. Supreme Court nullified the New York law regulating bakers' hours as "mere meddlesome interferences with the right of the individual."[15]

Women's work was lower paid than men's and sometimes more dangerous, particularly in the garment industry sweatshops. On March 25, 1911, the upper three stories of a ten-story building in New York City, where the Triangle Shirtwaist Company was located, caught fire. Most of the doors were blocked or locked, or opened inward, trapping the workers inside. There were no fire escapes. One hundred and forty-five of the five hundred employees, mostly young Jewish and Italian immigrant women, were burned to death or died jumping from the building. The New York State Factory Investigating Commission was formed in the aftermath of the fire.[16]

Children often did dirty work now done by machines. An example was the "coal breaker," inside of which twelve-year-olds crouched to clean slate and other refuse from the coal. Their pay was sixty cents for a ten-hour day. In the process their hands often became mangled and their backs permanently hunched like old men's. One reporter tried it out:

> Within the breaker there was blackness, clouds of deadly dust enfolded everything, the harsh, grinding roar of the machinery and the ceaseless rushing of coal through the chutes filled the ears. I tried to pick out the pieces of slate . . . my hands were bruised and cut in a few minutes; I was covered from head to foot with coal dust and for many hours afterwards I was expectorating some of the small particles . . . I had swallowed.[17]

William Z. Foster, leader of the great steel strike of 1919 and an early leader of the U.S. Communist Party, almost lost his life to industry as a young man:

> From 1901 to 1904 my revolutionary development suffered a rude interruption. The two and a half years I had worked with

lead in the type foundry as a child worker had undermined my health. The three years following in the fertilizer industry, where we usually toiled totally unprotected, in dense clouds of poisonous dusts, broke me so that the doctors pronounced me a consumptive. I was in a fair way to go to an untimely grave, grinding out profits for employers, as vast armies of workers had done before me. So I quit my job, pulled up stakes and headed for the West. I hoboed my way.[18]

Union-Busting and Working Conditions in Steel

From 1890 through 1910 the steel industry was the most dynamic sector of the economy. What happened in steel had a great effect on the rest of the economy, both because of the economic power it represented and by force of example. After 1890 the rapid reorganization of productive techniques, the huge investment in new equipment, and the gigantic increases in productivity per worker were accomplished only at a tremendous cost in human suffering. Steel owners found that they had to break worker control over the flow of production in order to implement their new plans, and they did so with a vengeance. Unions were smashed, labor activists purged, and the work week almost doubled in many cases.

The work pace, no longer under the control of skilled steelworkers, was speeded up mercilessly; as a result, injury and death rates in steel became a national scandal. Accidents were so frequent that they interfered with production, and owners feared that safety and health issues might become the basis for new union organizing drives. The response of the industry, led by the newly organized U.S. Steel Corporation, was to start implementing safety programs to reduce the number of accidents, and to use programs borrowed from Germany and Great Britain to compensate workers disabled on the job. Because these safety and compensation programs became the model for the rest of the nation, it is important to understand the situation of the steel industry around the turn of the century.

Before 1890 most steel was produced on contract by skilled steelworkers and laborers operating equipment owned by the capitalists. Workers would bargain with the millowners to decide what price they would receive, although their share could fall no lower than a certain amount per ton. This revenue was then divided up among the crew producing the steel, depending on their status within the production team. These arrangements were assured by the Amalgamated Association of Iron, Steel, and Tin Workers, a craft union, enrolling only skilled workers. The Amalgamated Association reached the height of its membership and power in the late 1880s and early 1890s, and the Homestead, Pennsylvania, mill was its biggest local. It was also the biggest mill of Carnegie Steel, the largest U.S. steelmaker.[10]

At Homestead about 25 percent of the 4,000 workers were in the union, where they controlled all aspects of production. This situation was unsatisfactory to Carnegie Steel for two reasons: according to the contract, workers got a constant share of the increasing sales of the mill, and they had the power to prevent the introduction of labor-saving technology if it did not benefit them. Carnegie resolved to break the Amalgamated Association's power at Homestead in a definitive test of strength. Before the contract expired in 1892, the management built a three-mile-long fence around the plant, with shoulder-level rifle holes every twenty-five feet. The workers were told that after June 24 Carnegie would deal with them only as individuals. On July 2, most unionized workers were locked out. The union, backed by all the workers, responded by shutting down the mill, renting a steamboat to patrol the Monongahela River, and organizing the whole town for resistance. When bargeloads of three hundred armed Pinkerton men were brought to force the mill open and bring in scabs, an open gun battle resulted: sixteen were killed, including seven Pinkerton agents, and the rest of the invaders surrendered to the workers and their families. They were beaten and run out of town. But after a four-month strike, Carnegie finally won the Homestead war with the help of the Pennsylvania militia, and the plant resumed production without the union. Loss of its most important outpost was the beginning of a rapid decline for the Amalgamated Association. After another disastrous strike in 1902, the union was completely uprooted from the steel industry in the United States.[20]

The steel bosses, who thereafter held total administrative control of steelmaking, set out to reorganize the industry as they wished. In the twenty years following 1892, productive techniques were revolutionized. Much physical labor was replaced by huge electrical cranes and intricate intraplant railway systems. Gigantic new furnaces and heaters were built, and formerly separate operations were integrated into single complexes. Production was increasingly carried out under orders, military style, according to a strict chain of command. Skilled workers who had been in charge of organizing production in their work gangs were demoted to semiskilled status at lower pay. Union activists were fired, and political conformity was enforced in small company towns by blacklisting known socialists and driving the rest underground.[21] Finally, the dirtiest and most unpleasant jobs were often filled by blacks and immigrants, which meant that these groups had higher casualty rates than native-born whites,[22] and which thus hindered labor's overall ability to unite against an increasingly centralized steel industry.

In awe-inspiring efforts to increase production, management speeded up the pace of work almost beyond belief. Production, stimulated by an apparently inexhaustible demand for steel, doubled every ten years or less. The work week, which had averaged eight hours per day six days a week, was increased to seven days a week for one-third of the U.S. Steel employees in blast furnaces and rolling mills. Twenty percent of those 153,000 employees worked an eighty-four-hour week.[23] The effect on working conditions was disastrous. In *The Steel Workers*, published in 1911, John A. Fitch wrote:

> There is always a fine dust in the air of a steel mill . . . not very noticeable at first, but after being in a mill or around the furnaces for a time, I always found my coat covered with minute shining grains. A visitor experiences no ill effect after a few hours in the

mill, but the steel workers notice it and they declare that it gives rise to throat trouble.[24]

Fitch thought that heavy physical work had been decreased by the new technology, but that the dangerous responsibilities for tons of molten steel and other materials had increased the "nervous strain" associated with being a steelworker. In most mills there were no "heat reducers," no fans, and no noise dampers. Men walked into the winter cold directly from the furnace heat, and washed up in troughs used to cool tools. A comparable study of British conditions at the time, published in 1902, showed that iron and steel workers had a "mortality figure 37 percent above that of the standard of occupied males."[25] Evidence collected from the death benefits fund of the Amalgamated Association of Iron, Steel, and Tin Workers suggested that injuries and "diseases caused by dust, heat conditions, and sudden changes of temperature [such as] tuberculosis and pneumonia" were important causes of death. But Fitch called attention to the twelve-hour day as "by far the greatest menace to health in the steel industry."

When the mills are running full the men are chronically tired. The upsetting of all the natural customs of life every second week when the men change to the night shift is in itself inimical to health. It takes until the end of the week, the men say, to grow sufficiently accustomed to the change to be able to sleep more than four or five hours during the day. And then they change back. . . . If this is true, it must be trebly so at the end of the twenty-four-hour shift, which is experienced fortnightly in Allegheny County (where Pittsburgh is located) by nearly 6,000 blast furnace men . . . it is a rare man who can keep his mental faculties alert . . . for twelve consecutive hours . . . in any accident case the long workday enters as a factor.[26]

As a result of the capital improvements, speedups, and long workdays, the steel industry was able to beat labor costs down from 22.5 to 16.5 percent of the total cost of making steel between 1890 and 1910.[27]

To counter the appeal of socialism and unionism, many steel companies instituted "welfare" programs to make life more pleasant for workers, without giving them any additional control over their lives. As Judge Gary, the head of U.S. Steel, told his chief executives: "Don't let the families go hungry or cold; give them playgrounds and parks and schools and churches, pure water to drink and recreation, treating the whole thing as a business proposition, drawing the line so that you are just and generous and yet . . . keeping the whole thing in your own hands."[28]

Under pressure from muckraking journalists, rising lawsuit costs, worker discontent over working conditions, and the costs of lost production, the steel industry led the way in the "voluntary" safety movement, financed and controlled by industry, with no outside intervention from government or labor. As early as 1907, U.S. Steel began to look for ways to reduce its casualty rates. The approach combined impressive engineering hardware with slogans and some on-the-job instruction in safety techniques. A glance at books such as George Alvin Cowee's *Practical Safety Methods and Devices*, published in 1916, constitutes persuasive evidence of the ingenuity of early safety engineers, and of the large resources placed at their disposal by the steel companies. Not surprisingly in this nonunion industry, the role of the steelworker in safety was a passive one, largely limited to looking at occasional safety posters and handbills, listening to rare safety lectures, and obeying regulations laid down by management. The effect of this highly praised safety "education" was to blame workers and absolve the company of responsibility for accidents.[29] The following example is taken from a safety publication of the time: "A workman in a foundry was wheeling a barrow and while passing by a crane the chain broke, the load dropped on him and he was killed."[30] The publication called this a clear case of "disobeying safety orders," rather than of equipment failure, even though existing knowledge showed conclusively that "there are defects in structure and use of chains which can be removed by engineering attention."[31] In its safety literature and studies, the steel industry paid no attention to speedup and long hours as possible accident causes, and the twelve-hour day in steel was only ended in the early 1920s.

In 1910 U.S. Steel inaugurated the Voluntary Accident Relief Plan, based on models developed in Bismarckian Germany by a conservative capitalist class under challenge from the fastest growing socialist movement in Europe.[32] This program, soon superseded by state workers' compensation laws, was "the first of its kind in the United States," and paid workers or their families fixed amounts for job-related injuries causing disability or death. The plan, for all its purported liberality, stated explicitly: "No relief will be paid to any employee or his family if suit is brought against the company," and workers who received "relief" from the plan were required to sign away any further rights to sue U.S. Steel.[33] U.S. Steel's experiments in safety and compensation became models for the rest of the United States.

The Origins of Workers' Compensation and the Compensation-Safety Apparatus

The compensation-safety apparatus is the complex of mostly private, corporate-dominated organizations which are concerned with compensation, workplace inspection, standards-setting, research, and education in occupational health and safety. It is called the *compensation-safety* apparatus because it emphasizes compensation over prevention and safety over health in its activities. It is an *apparatus* because it has executed the policies of business and insurance interests for decades as the only organized constituency in occupational safety and health. Only in the last ten years has its dominance been challenged by workers, unions, and their progressive allies.

Although Massachusetts passed the first factory inspection law in 1867, and Illinois passed one in the 1890s, these pioneering efforts were in fact unenforceable.[34] For example, "In Illinois, when Mrs. Florence Kelley was appointed factory inspector by Governor John P. Altgeld, not a lawyer or prosecuting attorney would handle the violations that she quickly detected. It was necessary for her to study law, gain admittance to the bar, and try such cases herself."[35] The issues of safety and compensation did not reach widespread legislative expression until big business felt sufficiently pressured to create its own policies for dealing with the problem and quieting the public outcry over carnage in the workplace.

At the turn of the century new business-sponsored organizations sprang up to help coordinate political activities, deal with labor, mold public opinion, and plan for the future. The National Civic Federation (NCF) was organized in 1900, and the American Association for Labor Legislation (AALL) in 1906. Their principal support came from the large corporations and banks. The policies of the NCF and the AALL often opposed those of smaller producers in the competitive sector, whose interests were represented by the National Association of Manufacturers (NAM). The first NCF president was "Dollar Mark" Hanna, wealthy banker and mine-owner, U.S. senator, and, in 1896, manager of William McKinley's successful Republican presidential campaign against William Jennings Bryan. The NCF also gained solid backing from such men as Andrew Carnegie and Judge Elbert H. Gary from the steel industry and Cyrus McCormack and George W. Perkins of International Harvester. Of the biggest capitalists, only the Rockefellers stayed aloof from active participation.[36]

In the first five or six years of its life the NCF preached the mutuality of interests of capital and labor and the importance of dealing with organized labor through collective bargaining and a written contract. Theorizing that there was no such thing as class conflict, the NCF tried to make arbitration a substitute for strikes in those places that unions did manage to organize. Although most of its chief corporate supporters bitterly resisted unions in their own factories, the NCF was, in principle, pro-union. In practice this meant trying to "channelize the labor movement into conservative avenues" wherever possible.[37] The NAM opposed this theoretical acceptance of labor unions, as they later initially opposed workers' compensation laws.

In the late 1890s Samuel Gompers, founder of the American Federation of Labor, decided that trade unions had to work with the large corporations, and he eventually became a vice-president of the NCF.[38] Both he and John Mitchell of the United Mine Workers were personal friends of the big-business leaders of the NCF. They hunted and dined with them and consulted them on investments.

The American Association for Labor Legislation (AALL) was active in promoting uniform legislation on a state-by-state basis throughout the United States, on the theory that this would prevent companies from moving their operations to states with less restrictive laws. The AALL made particular efforts in the areas of industrial diseases, industrial accident inspection, and workers' compensation. In dealing with industrial diseases, its major success was passing federal legislation banning the use of white phosphorus in making matches, thus eliminating "phossy jaw," a disease which caused the victims' jaws to stink and rot away.[39] But that success proved impossible to repeat on a broad scale.

Individual crusaders such as Dr. Alice Hamilton, the founder of occupational medicine in the United States, found themselves increasingly isolated and their work largely unrecognized. Hamilton's heroic achievements were due more to her personal knowledge and prestige than to the creation of strong institutions which could carry her work forward. Her famed "Illinois Survey" of health conditions in the lead industry was carried out on a "completely informal" basis with "no authority to enter workplaces." Since her survey could not mention dangerous workplaces by name and since her recommendations were unenforceable, she "made it a rule to try to bring before the responsible man at the top the dangers [she] had discovered in his plant and to persuade him to take the simple steps . . . that were needed." Where top management was uninterested in cleaning up, there was no recourse.[40] Those state laws regarding industrial inspection and occupational diseases passed in the first two decades of this century proved almost universally ineffectual in preventing accidents or industrial disease (see chapter 3).[41]

In addition to advocating specific programs for companies

and governments at the state and federal level, both the NCF and the AALL were concerned with more general "ideological and social problems," especially the threat of socialism, which was understood as the "only serious ideological alternative to . . . policies of social responsibility."[42] A long letter from an executive of the Lackawanna Steel Company to his company's vice-president in 1912 suggested that he contribute to the NCF because "the socialists and extreme radicals are very distrustful of it"; because it gave lawmakers "a sought-for excuse" to resist anti-business legislation; and because it projected an aura of civic-minded impartiality onto its programs. Though it sometimes looked as though the NCF got involved in programs which many employers would criticize, the executive wrote that it "only takes up a subject after it has assumed an important national aspect and it appears . . . that it . . . will be fought to an unfair conclusion. . . . Also there are many big men on the inside of the [NCF] who are able to inform and influence the action of legislators even after the [NCF] has been obliged to yield to popular clamor and let some subjects get away from them."[43]

The attitudes and actions of Theodore Roosevelt, president from 1901 to 1909, well expressed the anxieties of the corporate elite which shaped the country's legal and private response to the unrest over job health and safety. As New York City's police commissioner and as governor of New York, his first impulse had been to use troops against workers during strikes. Roosevelt considered socialism the most serious threat to capitalism, and fretted about how muckraking literature was "building up a revolutionary feeling." His closest presidential advisers were "almost exclusively representatives of industrial and finance capital," and he believed in the benefits of monopoly. His artfully created, totally undeserved reputation as a trust buster was used in key crises to convince the public of the government's impartiality. Once he understood the strength of the union in the anthracite coal strike of 1902, its public backing, and the disorder and "socialistic action" which could result from an attempt to break it, he heeded Mark Hanna and J. P. Morgan's call for mediation: "I was anxious to save the great coal operators and all of the class of big propertied men, of which they were members, from the dreadful punishment which their own folly would have brought on them if I had not acted."[44] The image of a president who could take "strong and independent" actions was of the highest symbolic importance in conferring legitimacy on the new economic order being brought about by the large corporations.[45]

Public consciousness of workplace casualties in the giant monopolies was a constant reproach to the consensus of "thoughtful men of all classes" which business was trying to forge. The popular and socialist media were full of stories about atrocities at work, which directly or by implication questioned the legitimacy of capitalism. Novels such as Upton Sinclair's *The Jungle* (1906) aroused public sympathy for workers, socialists, and unions, where none had existed before. A third of the members of the American Federation of Labor (AFL) were socialists[46] and unions, beaten down by the NAM's open-shop drive, only regained their 1904 membership levels in 1910. Thousands of workers joined such radical unions as the Western Federation of Miners, or took part in strikes led by the "Wobblies."[47]

By 1908 workers' compensation and job accidents had become major items on corporate agendas. Existing common law doctrines made it almost impossible for workers to collect damages for injuries suffered on the job because the worker had to prove the employer was at fault. This was particularly difficult for severely injured workers or for the families of workers killed on the job, who had to depend on the testimony of supervisors or coworkers subject to employer pressure. The boss could argue in self defense that (a) the accident had resulted from the worker's own carelessness; (b) the worker "assumed the risks" of the job in taking it; or (c) a fellow employee of the injured worker had caused the accident. This was usually enough to prevent successful worker suits.[48]

In some states the reaction of reformers was to pass laws weakening the common law defense of employers. In states where these "employers' liability" laws were passed, workers began to win more and larger settlements. As costs spiraled and open jury trials became a public embarrassment, the NCF began to lobby for a workers' compensation system which would "substitute a fixed, but limited charge for a variable, potentially ruinous one."[49] Soon corporate and government money for studies began to flow. The federal Bureau of Labor began to make careful estimates of the number of industrial casualties, and the new $10 million Russell Sage Foundation financed the Pittsburgh Survey, which paid a great deal of attention to working conditions. A whole book in the Pittsburgh Survey, Crystal Eastman's *Work-Accidents and the Law* (1910), was devoted to the wrongs of common law and "employers' liability" practices and the necessity for replacing them with a system of workers' compensation. The NCF led the forces for reform.

At the 1908 meeting of the National Civic Federation, speakers on workers' compensation included leading bankers, lawyers, insurance company executives, and Russell Sage Foundation experts—everyone but workers and their unions. By 1909 the NCF's Department of Compensation for Industrial Accidents and Their Prevention had become a center for lobbying and publicity. Experts from England and Germany (including a Major Piorkowski from Krupp) came to address the federation, and model compensation laws written by both the NCF and the AALL began to be requested by states all over the country. By the time government and foundation studies were completed and groups such as the NCF and the AALL had finished their conferences and come to their conclusions, the only question left was whether or not compensation insurance would be carried by private or state-run companies. Theodore Roosevelt's address to the NCF in 1911 included workers' compensation as a major theme. Conservative court opposition was overcome within the next year, and with active promotion by the NCF's reorganized Compensation Department and the AALL, by 1920 all but six states had some kind of workers' compensation law. Occupational diseases, however, never merited much attention in either the model or the actual laws, and so it remained until the late 1960s (see chapter 3).[50]

The new private workers' compensation systems and the unenforced industrial safety laws proved to be everything their corporate sponsors had hoped for. Both management and insurance interests benefited by the shift from chancy jury trials to administrative agencies whose employees could be bought off or coopted. Since it took a long time for a group of expert plaintiff lawyers to develop, the next two decades were "largely a period of unilateral advocacy" by an "expert, specialized, full-time . . . bar" that defended only the interests of the insurance companies and corporations.[51] Amendments to compensation laws by state legislatures followed the same pattern: costs to companies were stable and averaged around 1 percent of payroll, occupational disease payments were almost nonexistent, and companies were protected from negligence suits by common law. Physicians were hired to deal with work injuries and to represent employers within the compensation bureaucracy, creating that peculiar institutional ghetto called "industrial medicine."[52]

The corporate response to the problem of preventing injuries took two forms: the founding of the "voluntary" industrial safety movement and the passage of ineffective and unenforced industrial inspection laws. The first was by far the more important; its doctrines and practices, under the leadership of the National Safety Council, were to dominate the field of occupational safety and health for over half a century.

The organization which would become the National Safety Council was conceived in 1911 at a meeting of the Association of Iron and Steel Electrical Engineers, and was formalized the next year in Milwaukee to the incantations of Dr. Edwin A. Steiner, professor of Applied Christianity at Grinnell College. Judge Gary, of U.S. Steel and the NCF, was a guiding light of the new group, and the sixteen charter members of the Committee on Permanent Organization were rather broad-based: five from federal agencies; four from the steel industry; two each from state government and the NAM; and one each representing the railroads, the insurance industry, and the Red Cross. No unions were involved.[53]

The National Safety Council was conceived of as a "separate safety organization, national in scope, that could better serve as a coordinating agency and general clearing house for all phases of accident prevention." Several features of the steel industry's experience became permanent aspects of the voluntary safety movement: it was undertaken without government compulsion; accidents were blamed largely on worker carelessness; workers and unions were left out of decision-making (except for some later tokenism); occupational diseases were ignored or suppressed from public view; and speedup and long working hours were considered to be outside the realm of industrial health and safety.[54]

This corporate-financed and -controlled compensation-safety apparatus dominated occupational safety and health policy without serious dissent from the early 1900s until the late 1960s. In 1926 a major summing up of corporate experience in the field (in which unions were barely mentioned) revealed a self-satisfied tone and a sense that the main challenges had been met and overcome and that the field had nothing new to offer.[55] The silly safety posters workers have come to expect today in place of a real program began to make their appearance. One plant newspaper sponsored a "safety first limerick contest." The winning entry, accompanied by a cartoon of a guy heels over head in the air, went like this:

> A grouchy guy, Isaac Maloney
> Said "Safety first—bah—that's boloney!"
> 'Til he once looped the loop
> When he stepped on a hoop
> *'Tis "Ikee" not "safetee" that's phoney.*

In "putting safety across to the worker," safety contests took on the form of athletic events, with injuries a test of team spirit. Here is how one oil industry editor put it:

A young man in one of the large industrial plants in Philadelphia had the end of his finger torn off . . . the other day. His department had been striving for a 100 per cent no-lost-time record for that month, and up to his injury its record was perfect. Although his accident was mighty painful, and he was advised to go home and remain there for a few days after the doctor dressed it, this young man was so anxious to keep that 100 per cent record perfect that he refused to leave, and insisted on going on with his work. The *morale* of his department did it.[56]

An increasing emphasis on the psychological defects of workers made it seem as if the safety establishment were beginning to believe its own propaganda. Studies by the British Industrial Fatigue Research Board made during World War I became the basis for a whole new area of research. The "accident-proneness" concept, coined in 1926, was picked up by a whole wave of industrial psychologists in the United States, who searched for the causes of accidents in the characteristics of individual workers. Although some studies examined such factors as lighting and rate of production, most blamed the victim of the accident rather than seeking to modify the conditions which produced it. If accidents were caused by worker "carelessness," "light-heartedness," "excitability," "psychomotor retardation," or "deficiency of English," why bother about workplace design?

As late as 1966, the *American Handbook of Psychiatry*'s article on "Industrial and Occupational Psychiatry" gave "examples of a variety of syndromes and situations which may merit the attention of a psychiatrist."

Occupational Syndrome	*Clinical or Dynamic Diagnosis Associated*
accident syndrome	impulsive character, anxiety reaction
moonlighting	compulsive personality, often with marital problems
pulmonary insufficiency ("pneumoconiosis," "emphysema," "chronic bronchitis")	depressive reaction, anxiety reaction, psychophysiological reaction (asthma)
women employees	physiological cycles
grievance proneness	paranoid personality, compulsive personality, depressive reaction[57]

The authors make no plea for cautious examination of the real conditions of work before making use of their outlandish labels—apparently there were no problems that could not be traced to the mental disease or personality defects of workers if a clever psychiatrist looked hard enough.

The nitpicking approach of the accident-proneness school was in sharp contrast to the boldness characteristic of early engineering work in industrial safety which focused on redesigning and guarding machines and on regularizing and smoothing the traffic of workers and materials. The changing emphasis exemplified the decline of corporate interest once the obvious problems which cost corporations a great deal of money in lost production and compensation were solved. Sincere occupational safety and health specialists were caught between their desire actively to promote better conditions and a lack of corporate interest. One president of a large corporation told his "safety man": "You know I am very busy and have little time to give to safety, but keep me in touch with your work. It is a non-controversial subject. I can make train conversation of it."[58]

In the absence of strong unions or a national movement around health and safety, progressive specialists had nowhere to go outside the corporations. They were left to plead for recognition of the importance of their mission "from the men that control the destinies of business life." In 1926 W. H. Cameron, director of the National Safety Council and an astute analyst of the decline of the corporate-backed safety establishment, wrote: "It must . . . be admitted that there is a lack of leadership in the safety movement. Because the economic stake in many industries is small, small men are given the responsibility. The minds of the major executives are occupied by the major problems of production, distribution, finance."[59] By the mid-1920s the big names from government and business were no longer thinking and speaking and writing about occupational safety and workers' compensation. With outside agitation under control, the field was turned over to the less well trained specialists of the compensation-safety apparatus.

Attitudes of Unions and Workers

The struggle for humane working conditions has taken many forms in its long history. When the first unions were organized in the United States they brought pressures to shorten the work day. In June 1835, two years after its founding, the 10,000 workers of the Trades' Union of the City and County of Philadelphia organized what was probably the first successful general strike in America. The shoemakers, followed by the building trades workers, cigar-makers, carters, saddlers and harness makers, smiths, plumbers, bakers, printers, and even by the unskilled dock workers, struck for a ten-hour day. After a mass meeting of citizens endorsed the union's demands, the city council agreed to a ten-hour day "for all municipal employees." The demand swept through New York City, Boston, Baltimore, Washington, D.C., Newark, St. Louis, and other cities, until the collapse of the unions in the panic of 1837. In 1847, after agitation from labor, New Hampshire became the first state to pass a law making ten hours the legal work day, "unless otherwise agreed to by the parties." New Hampshire's lawmaking lead was followed by Pennsylvania, New Jersey, Ohio, and Rhode Island by 1853, but enforcement was another matter.[60]

Such struggles were isolated, however, and unions did not become common until the growth of the Knights of Labor in the 1880s. Agitation and organization for the eight-hour day was the basic issue in the mass strikes begun on May 1, 1886, which brought out 80,000 strikers in Chicago and scores of thousands in other cities, memorialized by May Day, the international workers' holiday. The Haymarket bombing in Chicago three days later and the subsequent show trials constituted the employers' mostly successful attempt to destroy the eight-hour movement and the union organizing that proceeded around it.[61]

In the earliest period of lobbying and public relations around workers' compensation, organized labor was conspicuous by its absence. In 1860 unions had less than two million members, and many of them were fighting for their very survival.[62] As unions were liquidated from the steel industry after the Homestead strike, the work week for some workers nearly doubled, to eighty-four hours. Unions still had had little success in penetrating meatpacking, oil, and most of the other big new industries. Clearly, the first task for workers who wanted to reduce working hours and improve working conditions was to organize.

Union reluctance to join the NCF's rush to write compensation laws is understandable. After all, the basic models for workers' compensation and industrial safety in the United States were first tried out by the steel industry, in a context of union busting and a corporate welfare paternalism designed to keep unions out. Furthermore, the compensation programs were being propounded at a time when workers were beginning to win larger and more frequent negligence judgments against employers, as traditional common law defenses were weakened and juries became more sympathetic. James Weinstein has pointed out that:

> Compensation laws, in contrast, could be expected only to pension off the worker during his period of disablement at something less than his regular wages. In addition, almost all unionists, conservative or socialist, opposed government regulation of working conditions on the theory, often only implicit, that government was controlled by business, either directly or through conservative politicians and judges.[63]

Later, when Samuel Gompers reluctantly endorsed the concept of workers' compensation under pressure from his big-business friends and from a belief in the inevitability of some kind of legislation, the labor movement and the socialists each developed their own positions on the issue. But business domination of the political process in a climate basically favorable to private enterprise made it difficult for either to come up with workable alternatives. The programs of both labor and the socialists called for full compensation of lost wages (as opposed to one-half or two-thirds); retention of the right to sue at common law before a jury; and state-owned insurance companies, to prevent the diversion of most premiums to insurance interests rather than injured workers. Crystal Eastman, for example, had found in 1910 that only 20 to 37 percent of employer premiums in employers' liability were paid out as benefits in some form by the insurance companies.[64] With the addition of reasonable presumptions concerning occupational disease liability, the labor/socialist proposals of sixty years ago would make a fine platform for a radical restructuring of workers' compensation today.

As labor and the socialists had feared, almost none of their demands were won. In no state were benefits close to 100 percent of lost wages. In New York State, organized labor's program banning private insurance companies was sacrificed for fairly high initial benefits, and a state-owned company was allowed to coexist with private companies. In Missouri the institution of workers' compensation was held up by a strong labor movement until 1926 over the issue of a state fund, but the battle was lost. Of the big industrial states, only Ohio totally excluded private insurance companies, and today Ohio pays out 96 percent of its premium income to victims and doctors, as opposed to 53 percent for private insurance companies nationally.[65]

The Depression sapped interest in worker health and safety; people were desperate for work. Even the mass slaughter of workers during the building of a hydroelectric tunnel near Gauley Bridge, West Virginia, to bring power to a Union-Carbide subsidiary, failed to awaken public interest, although congressional hearings on the disaster were held.

The workforce at Gauley Bridge was made up of poor people, mostly blacks, who flocked to the jobs at fifty cents (later thirty cents) per hour. Workers and their families were herded into shacks no larger than chicken coops, with twenty-five to thirty people per room. The water tunnel was being cut through almost pure silica, and the dust was so thick that workers sometimes could see barely ten feet in the train headlights. Instead of waiting thirty minutes after blasting, as required by state law, workers were herded back into the tunnel immediately, often beaten by foremen with pick handles. According to one report:

Increasing numbers of workers became progressively shorter of breath and then dropped dead. Rhinehart-Dennis contracted with a local undertaker to bury the blacks in a field at fifty-five dollars per corpse. Three hours was the standard elapsed time between death in the tunnel and burial. In this way, the company avoided the formalities of an autopsy and death certificate. It was estimated that 169 blacks ended up in the field, two or three to a hole. . . .

Toward the end of the project, some workers bought their own respirators for $2.50. The purchasing agent for Rhinehart-Dennis was overheard to say to a respirator salesman, "I wouldn't give $2.50 for all the niggers on the job." The paymaster was also heard to say, "I knew they was going to kill those niggers within five years, but I didn't know they was going to kill them so quick."[66]

Altogether, over 470 men died and 1,500 were disabled. Years later, when survivors sued Rhinehart-Dennis for negligence under common law (silicosis wasn't compensable), rumors of jury tampering were widespread, and a group of 167 suits were settled out of court for $130,000, with half going to the workers' attorneys. Blacks received from $80 to $250 and whites from $250 to $1,000 each. Later it was discovered that one of the workers' law firms had accepted a $20,000 side payment from the companies.[67]

Yet even this monstrous episode failed to catalyze reformist energy in the area of health and safety. Shortly after the adjournment of the Gauley Bridge hearings in 1936, a group of large industrialists and scientists met under the auspices of the Mellon Institute in Pittsburgh to discuss the problem. Out of these meetings was formed the Air Hygiene Foundation, later renamed the Industrial Health Foundation (IHF). Its publicly announced purpose was to promote "better dust conditions in the industries," but its real purpose was to forestall massive claims. The confidential report of the proceedings noted:

Because of recent misleading publicity about silicosis and the appointment of a Congressional committee to hold public hear-

ings, the attention of much of the entire country has been focused on silicosis. It is more than probable that this publicity will
result in a flood of claims, whether justified or unjustified, and
will tend toward improperly considered proposals for legislation.[68]

One of the first programs of this foundation, a public relations campaign to "give everyone concerned an undistorted
picture of the subject," was almost wholly successful. Occupational disease claims hardly rose at all, and the foundation,
under its different names, became a permanent and important part of the compensation-safety apparatus, special-
izing in debunking the seriousness of occupational health
problems. By 1940 there were 225 member companies,
including many corporate giants. At the IHF's fifth annual
meeting, the chairman of the membership committee made
a pitch in the following terms:

A survey report from an outside, independent agency carries
more weight in court or before a compensation commission than
does a report prepared by your own people. One of the brilliant
features of IHF is this; it is a *voluntary* undertaking by industry
to protect industrial health. And where industry attacks a great
social-economic problem voluntarily, there is no necessity for
government to step in and regulate.[69]

The IHF later branched into other areas of occupational
health, successfully providing medical cover for many other
industries, until its credibility began to be questioned by the
new activists of the last decade.

Despite the lack of a national movement around health and
safety after 1920, working conditions were sometimes prime
causes of rank-and-file unrest. In 1935, 108 black steelworkers in northern Indiana sued subsidiaries of U.S. Steel
for failing to provide healthful working conditions. The
worker-plaintiffs were mostly furnace cleaners and coke-oven
workers who, like immigrants before them, had been given
the dirtiest jobs in the mills. They charged that their jobs had
caused tuberculosis, silicosis, and other lung diseases. The suit

was settled out of court for an undisclosed amount in 1938.[70]

Union organizer Stella Nowicki tells of one incident in the 1930s in Chicago where safety triggered a union organizing drive. She was canning meat (and cockroaches) in an atmosphere so "hot and steamy" that women used to pass out repeatedly.

> The thing that precipitated it is that on the floor below they used to make hotdogs and one of the women, in putting meat into the chopper, got her fingers caught. There were no safety guards. Her fingers got into the hotdogs and they were chopped off. It was just horrible.
>
> Three of us "colonizers" had a meeting during our break and decided that this was the time to have a stoppage and we did. (Colonizers were people sent by the [Young Communist League] or [Communist Party] into points of industrial concentration that the CP had designated.... The colonizers were like red missionaries. They were expected to do everything possible to keep jobs and organize for many years. All six floors went on strike. We said, "Sit, stop." And we had a sit-down. We just stopped working right inside the building, protesting the speed and the unsafe conditions. We thought that people's fingers shouldn't go into the machine.... The women got interested in the union.
>
> We got the company to put in safety devices. Soon after the work stoppage the supervisors were looking for the leaders because people were talking up the action. They found out who was involved and we were all fired. I was blacklisted.[71]

The labor shortage during World War II bought a huge influx of workers into new jobs and led to much higher accident rates. For management it "forcibly brought to light the importance of conserving manpower in order that production . . . be maintained."[72] For the first time federal grant support went to state occupational safety and health programs, some of them maintained as late as 1950. In 1951 Senator Hubert H. Humphrey introduced a bill proposing that the federal government write and enforce safety standards, and other bills were introduced that year (and again in

1962) to procure aid for the states to promote job accident prevention, but these were unsuccessful and never attracted much attention.

In 1952 a seven-month strike over a sandlike dust which produced a lung disease similar to silicosis at a mine and mill in Lompoc County, California, failed to incite national interest in worker health on the job, though it received much publicity in the California press and was successful in attaining many of its objectives.[73] In the 1950s the Ladish Blacksmiths and Boilermakers Local 1509 made a major effort to secure routine compensation for severe occupational hearing loss in Wisconsin. They were unsuccessful, and were so disappointed with the lack of support on the noise issue that they quit the Wisconsin AFL-CIO in protest.[74] Carl Carlson, head of the safety committee of United Auto Workers Local 6 at International Harvester near Chicago, began investigating the problem of noise in 1959, but in the absence of a wider movement, progress was slow. Fifty years of domination of occupational safety and health by large corporations and insurance companies had made it almost impossible for workers to find sympathetic specialists or a suitable climate of opinion for action.

The New Movement in Health and Safety

Serious attack on the compensation-safety apparatus began in the late 1960s in the coal mines, over the issues of black lung and safety. The issue was taken up by workers and unions and their allies in industries throughout the United States.

In the mid-1960s coalminers began to fight for state laws granting them workers' compensation benefits for disability caused by black lung. Actively opposed by the leadership of the United Mine Workers, they founded their own organizations—the Black Lung Associations and the Association of Disabled Miners and Widows—aided by sympathetic physicians, lawyers, and other specialists. In February 1969, 42,000 of West Virginia's 44,000 coalminers carried out a wildcat strike for three weeks, and marched on the state capital to get a black lung compensation bill passed. The miners' organizing and agitation led to congressional passage of the Coal Mine Health and Safety Act of 1969 and the Black Lung Benefits Act of 1972. The nucleus of people who worked on health and safety issues eventually combined with the leadership of Miners for Democracy, a rank-and-file group which took control of the leadership of the United Mine Workers from the corrupt and murderous Tony Boyle clique.[75]

In 1964 the working conditions issue outside the mines was first brought up in a new forum at a President's Conference on Occupational Safety, without specific legislative proposals in mind. In 1965 the Department of Health, Education, and Welfare published a report recommending vastly increased expenditures to deal with the multitudinous new health hazards, but there was as yet little interest in the issue from the unions, natural allies of such a proposal. Finally in 1968 President Lyndon B. Johnson's office introduced a bill fashioned by reformers both within and outside the government as a way to call attention to the issue. Management testimony opposed the bill, while initial union reaction was favorable, if desultory. But interest generated by the 1968 hearings, the environmental movement, and the high attention paid to coal mine conditions (renewed, typically, by an explosion which killed seventy-eight miners) kept the issue alive until 1969, when Richard M. Nixon became president.[76]

The most lucid support for a strong law came from the Oil, Chemical and Atomic Workers, Ralph Nader and his supporters, the Steelworkers, the AFL-CIO, the United Auto Workers, and a few senators and members of Congress. The hearings show that groups from the compensation-safety apparatus (including the National Safety Council and the organized representatives of company medicine, safety engineering, and industrial hygiene) followed the line of the National Association of Manufacturers and the U.S. Chamber of Commerce in backing weak bills proposed by the Nixon administration. Given the lukewarm support from the top leadership of most unions, the wonder is that any legislation passed at all. The authors of *Bitter Wages* (Ralph Nader's Study Group Report on Disease and Injury on the Job) recount that "the most surprising source of apparent support came from the Nixon Administration's attempt to grab hold of the safety and health issue. The Republicans were making a loud pitch for blue-collar support and had unleashed a torrent of rhetoric concerning the 'silent majority' and the 'forgotten Americans.' "[77]

Nixon finally signed the Occupational Safety and Health Act of 1970 into law on December 29, 1970. The law promises much more than it has delivered. Employers are boldly required to provide a workplace "free from recognized hazards that are causing or are likely to cause death or serious physical harm to employees" and to meet the specific standards promulgated by the Labor Department's Occupational Safety and Health Administration (OSHA). To guarantee compliance, the law gives OSHA the power to inspect workplaces, make citations for violations, and propose penalties. In situations of "imminent danger," the Labor Department is authorized to seek an injunction in federal court to shut down the offending operation. The law makes explicit provision for workers to call in inspectors without giving employers advance warning. Workers have the right to a "walk-around" with OSHA compliance officers and to point out suspected violations. Anonymity is guaranteed (if requested) to the worker making a complaint, and workers are supposed to be protected against employer reprisals.

Standards are set by the OSHA administration with the advice of the National Institute for Occupational Safety and Health (NIOSH). All the early standards for safety and health were adopted directly from private standards created by such industry-supported organizations as the American National Standards Institute, and the setting of new standards has proceeded extremely slowly because of deliberate OSHA obstructionism. A memo from OSHA's chief administrator to Nixon's 1972 election staff recommended that the promulgation of "highly controversial standards" (i.e., for cotton dust, etc.) be avoided, and that "four more years of properly managed OSHA" should be used "as a sales point for fund raising and general support by employers."[78]

Five years of OSHA have led to few improvements in working conditions: only a small proportion of worksites are inspected every year, and inspections are fairly common only in workplaces of at least 500 workers. Moreover, penalties for violation of the standards are rather small: "From its inception in April 1971, through January 1975, OSHA made a total of 206,163 inspections resulting in 140,467 citations alleging 724,582 violations with proposed penalties totaling $18,186,627," an average of about $25 per violation. In the first eleven months of fiscal year 1975 OSHA inspections of some type were carried out in workplaces employing only 11.5 million workers out of a national total of over 80 million.[79] Appeal of penalties and work orders written by inspectors has become routine, especially for large companies, further slowing down the "enforcement" process. Successfully contesting employer appeals is bewilderingly complex for workers without a lawyer,[80] and only one union, the Oil, Chemical, and Atomic Workers Union, routinely contests such appeals. Workers' "rights" under the OSHA law are often so hedged with restrictions or so little known that they are impossible to exercise without support from a knowledgeable union staff, making them almost meaningless for most workers, particularly in the twenty-five states where enforcement has been returned to state governments.[81]

Thus, since the passage of the OSHA law in 1970, the compensation-safety apparatus, backed by the large corporations and a compliant administration, has been able to absorb and neutralize most threats to its dominance of the new federal regulatory apparatus. The teeth of OSHA have been pulled out through low enforcement budgets, restrictive regulations, appeals favoring employers, low fines, and the partially successful effort to return rule-making and inspection to the states. Away from the coal mines, health and safety has rarely escaped the bureaucracy to become a mass issue, except at isolated factories and mines. Workers, fearful for their jobs in a period of permanent high unemployment, have been reluctant to push too hard for improved working conditions. Even the victory of passing a stringent new standard for exposure to vinyl chloride was an ambiguous one. The battle for a strong standard, fought mostly within OSHA's labyrinthine bureaucracy and the mass media, successfully tied up a good share of the reformers' scientific and legal talent for months, and the outcome affected only a few thousand production workers. The slightly increased production costs of polyvinyl chloride plastic will be paid for by the public, since vinyl chloride is in high demand and its producers are able to "coordinate" their prices upward as their need arises.[82] Comparable new restrictions on the use of asbestos by small firms in the competitive sector, where union pressure could not be exercised, have been largely ignored.[83]

Given such lax enforcement and delay, the most important consequence of the OSHA law has been to place the issue of the work environment on the agenda for workers, unions, health specialists, and the general public. Unions, often under pressure from their memberships, are forming health and safety departments and teaching their members to identify and correct hazards. Workers and progressive scientists, lawyers, and health specialists are organizing and working together through grassroots health and safety coalitions, such as the Chicago Area Committee for Occupational Safety and Health, Urban Planning Aid (now MassCOSH) in Boston, and the Philadelphia Area Project on Occupational Safety and Health. Medical researchers, led by Doctors Irving Selikoff and Sam Epstein, have formed the Society for Occupational and Environmental Health, which admits nonprofessionals as members; and the general public's awareness of working conditions has increased through a few books and mass media programs. Within this context it is possible to point to specific advances for working people: the establishment of their rights under OSHA; the general acceptance of higher and more realistic estimates of the size of the problem; the new questioning of the traditional role of the company doctor; the new interest in occupational disease; the increased number of collective bargaining and research initiatives in health and safety; and coalfield gains regarding black lung compensation.

However, it is doubtful that the advances of the past decade can be preserved, let alone extended, in the context of a capitalist system where real wages have been stationary or declining,[84] the proportion of unemployed and marginalized workers continually creeps upward,[85] multinational firms increase their power over U.S. and foreign production, and thus their ability to lower costs by producing abroad or driving down U.S. production costs,[86] social and environmental program cutbacks are the order of the day,[87] the proportion of unionized workers is declining, and the posture of union leadership generally has been one of "public aggression and private cooperation."[88] Although some shops and industries will move against the general trend, there is no reason to believe that working conditions can improve if the overall standard of living of the working class is being pushed back; for the problem of job health and safety is only one issue among many with which workers must contend.

There is much evidence suggesting that corporate and governmental expenditures on working conditions have already peaked and begun to drop. Employers and their associations have come closer to gutting the OSHA law in Congress and have garnered support among President Carter's closest advisers. The innovative workers' compensation legislation proposed by Senators Jacob Javits and Harrison Williams in 1973 never passed, and newer versions have been successfully watered down. The insurance carriers around the country are mounting an offensive to make occupational diseases and other permanent disability benefits harder to collect, unions for the most part have not dedicated substantial resources to the problem, and hiring of health and safety staff has leveled off.[89] For lack of union support, many of the grassroots groups have died off. Finally, fundamental questions concerning the role of the private insurance industry in workers' compensation, or the right of workers to help design factories, shut down dangerous jobs, or choose their own doctors have hardly advanced beyond the threshold of concern, although they are crucial if the struggle is to advance. In a general atmosphere of political conservatism, a holding action seems to be the order of the day. What happens in the workplace cannot be separated from what happens in the rest of society.

|2|
The Official Body Count

The Vietnam war has taught us to question official casualty figures, and industrial casualty statistics deserve the same skepticism. Throughout corporate and political life the creation and manipulation of information in all forms is of tremendous importance. It is essential for controlling everyday decisions and influencing the very definition of modern problems. Accurate information is sometimes necessary to carry out corporate and government policy, but where the truth contradicts official preconceptions it is ignored or denied. In most cases the real decisions have been made beforehand, and the Ph.D.s and professors are trotted out, in the pay of the mighty, to provide their "expert" sanction to the proceedings.[1] Financial reporting procedures that mystify more than they reveal,[2] a Freedom of Information law enforced mostly for lobbies and corporations that already have the inside word,[3] and an OSHA data system that ignores most of the casualties it was ostensibly designed to record are commonplaces of bureaucratic life.

Although libraries, computers, and our own heads are stuffed with "information," we can't find out what we want to know. Despite the spectacular growth of industrial casualty reporting since the late 1960s, we still have no believable estimates of the total number of those sickened, injured, or killed as a result of industrial activity. To understand the reasons behind this constant cataloguing of misinformation, we must patiently examine who collects the statistics, what their methods are, and what use they make of the results.

Who's Counting the Casualties?
The Recorded Death Toll from Work Accidents

In 1908 Frederick L. Hoffman, writing for the *Bulletin of the U.S. Bureau of Labor*, asserted, "Thus far no national investigation of the subject of industrial accidents has been made to determine the true accident risk in industry, and the statistical data extant are more or less fragmentary and of only approximate value."[4] Nevertheless, he went on to imply, using evidence culled from death certificates, that 15,000 to 17,500 accidental deaths annually among male workers were "more or less the immediate result of dangerous industries or trades." In 1915 Hoffman upped his estimate for industrial accident deaths "conservatively" to 25,000 (apparently then including women), although he still bemoaned the "incompleteness" of the data.[5] After 1919 Labor Department estimates were based on the number of deaths reported by the workers' compensation systems that had been created in most states. The number of deaths recorded between 1919 and 1927 ranged from 9,392 to 12,531.[6] The change in the information source had led to a 50 percent drop in the reported total of industrial accident deaths.

In the late 1920s the National Safety Council (NSC) took over responsibility for estimating these figures, but *Accident Facts*, the NSC's widely circulated statistical handbook, gives almost no indication of where its figures come from. The reported totals for the years 1963 through 1971 oscillated narrowly between 14,100 and 14,500.[7] These results were given a spurious appearance of accuracy by the inclusion of annual percentage changes in the number of deaths. These concocted measures usually implied that the workplace was constantly becoming safer.[8]

Here is how President Tofany of the National Safety

Council described its procedures for estimation of the job accident death total in 1974:

> NSC statisticians use the official death certificate counts made by the vital statistics authorities in HEW (currently the National Center for Health Statistics) as the source for the numbers of work, home, and motor-vehicle deaths. Within the overall total of accidental deaths, those involving motor-vehicles are precisely tabulated. Such precise tabulations are not made, however, for each of the other three categories (home, work, public).[9]

Since death certificate counts reveal that less than half of the total of 115,000 accidental deaths are caused by motor vehicles,[10] it becomes crucial to find out how the NSC divides up the rest, given the inability of death certificate information to provide accurate answers:[11]

> Since every nonmotor-vehicle death must be classified into one of the three categories . . . this provides an effective cross check of the separate counts. If the work count is reduced by 3,000, then this number must be added to the home and public totals. This added check, plus dozens of separate inputs, plus years of testing and refining the estimating procedures, leads the NSC to believe that its count of 14,100 *total* work deaths is very close to the true number.[12]

The "dozens of separate inputs" referred to consist mostly of an incomplete and out-of-date set of state workers' compensation reports,[13] reports from an unrepresentative sample of NSC members (mostly large corporations), and reports from such federal agencies concerned with work accidents as the Bureau of Mines. A spokesman from the NSC statistical department admitted that there was no standard written procedure used to compute the industrial accident death total. "It's basically guesswork," he said, "nothing like what I learned in school; but you get used to it after you've been here awhile."[14]

Work Injury Statistics

Between 1919 and 1927 the Labor Department published estimates of the national incidence of accidental job injuries, based on data culled from the workers' compensation system. Reported injuries for those years ranged between 1,274,220 and 1,687,957.[15]

In 1920 the Bureau of Labor Statistics published a method they had developed for counting the incidence of industrial injuries in a given workplace, which was eventually adopted by the National Safety Council.[16] The evolution of this standard (now known as the American National Standard Z16.1 Method of Recording and Measuring Work Injury Experience) provides a good lesson in how the compensation-safety establishment has been able to take over the definition of problems to suit its interest in minimizing the industrial casualty problem. Although the standard was originally created by the Department of Labor's Bureau of Labor Statistics, in 1926 it was turned over to a private industry-controlled standards-setting group (the predecessor of the American National Standards Institute). Between then and 1967 it went through four revisions,[17] making it useless as a comparative measure of the number of work injuries. In the late 1950s Frank McElroy, chief of the Division of Industrial Hazards, Bureau of Labor Statistics, wrote as follows about those changes:

> In all honesty, we have to recognize that most of the specific rules introduced into Z.16.1 have the effect of reducing the range of reportable injuries. . . . If we accept this premise, as I feel we must, all of our statistical indications of improvement in the volume of work injuries become questionable. Have we really succeeded in bringing injury occurrence in manufacturing to the lowest level in history, or do our figures largely reflect shifts in reporting rather than substantive improvement? Are we, in effect, kidding ourselves?[18]

The Z16.1 standard records only "lost-time" accidents which cause death, permanent disability, or at least a full day's inability to work. Occupational diseases are almost never recorded, and as far as I know have never been listed as a separate category in statistics published by the National Safety Council or the Bureau of Labor Statistics. The recording provisions of the standard are widely ignored or circumvented in companies' public relations "safety" contests:

> As of January 17, 1970, the safety sign in the Rohm and Haas chemical plant at Pasadena, Texas, said that the plant had recorded 458 consecutive days without a lost-time accident. Asked if he thought the figure was distorted, union workmen's committeeman Bob Walter says, "Very much so. We've had the problem of people getting burns from steam hoses and . . . so forth and they haul them in there on crutches. When I first went to work out there I got a skin irritation from benzene. My hands swelled up. Oh, they were huge! Well, they brought me back out there and set me at the desk in front of the telephone and asked me to answer the phone for . . . about two weeks to keep from listing a lost-time accident. . . . They'll bring the people back in on stretchers if they have to."[19]

Even when the Z16.1 standard was correctly observed, according to the 1970 *Evaluation of the National Industrial Safety Statistics Program* known as the Gordon Report, it failed to record the 90 percent of "serious" but nondisabling injuries, such as eye injuries, loss of consciousness, fractures, hospitalization for observation, treatment by a physician, restriction of work or motion, or assignment to another regularly scheduled job because of a work injury. If the Gordon Report criteria for serious injuries had been used for statistical purposes, 25 million rather than 2.5 million job injuries would have been recorded each year by the National Safety Council.[20] Howard Pyle, then president of the National Safety Council, admitted in an interview that there were no better work injury figures than those contained in the Gordon Report;[21] but to this day the NSC continues to publish statistics according to the outdated Z16.1 standard. For the National Safety Council nothing has really changed.

The passage of OSHA led to some change in official record keeping. The current Labor Department definition of what constitutes an occupationally caused death or injury has been broadened to make many more injuries reportable. As a result, the first complete Labor Department survey under the new rules revealed that about 5.6 million employees were injured—in contrast to the 2.3 million estimated for 1971 under the National Safety Council's Z16.1 standard.[22] But even under the new rules employers are still solely responsible for reporting injuries and diseases incurred on the job, and workers have no right to check up on the accuracy of their reporting, unless this is specifically provided for by contract language. Workers are normally allowed to see only an annual summary (which is supposed to be posted during February), rather than the "employer's log" of injuries and diseases as they happen, or the "supplementary record" of workers' compensation cases and other information,[23] which together form the basis of reports. According to Carl Carlson, safety chairperson of the United Auto Workers Local 6 (Melrose Park, Illinois), the International Harvester Company is "doctoring the facts" about injuries and diseases in its annual survey, and NSC president Tofany's Senate hearings testimony indicates that massive employer under-reporting is general throughout the nation.[24] A study carried out since the passage of OSHA proved that over 40 percent of work injuries uncovered in a medical survey of workers in small plants had gone completely unreported, either in the employer's log or the supplementary record of workers' compensation cases:

> The answers on the questionnaire indicate that a large body of unreported occupational accidents may exist. The Employer's Log was not available at 22 of the 64 establishments in the medical survey. The excuse given was that there were no work related injuries or illnesses to report. Workmen's Compensation claims were found for workers in 10 of the 22 establishments without Employers' Logs.[25]

Clearly, any system based on employer self-reporting will understate the extent of the problem, since it is hardly in employers' interests to keep accident records which will reflect badly on themselves or on employers as a class. Nevertheless, the head of OSHA record keeping finds he has "no reason to believe" that employers "consciously provide inaccurate information."[26]

Occupational Disease Rates

Early attempts to find out the incidence of occupational disease included the appointment of study commissions (such as Dr. Alice Hamilton's famous Illinois Survey), and the passage in the first two decades of this century of state laws which required the reporting of diseases to state health departments.[27] Often, laws were passed requiring that workers in particularly dangerous jobs be examined periodically by a physician to determine if they were getting sick, with the reports to be forwarded to state health departments. As these reporting and examination laws were largely ignored, however,[28] the only sources of occupational disease statistics were the state workers' compensation agencies. Since occupational disease claims (except recently in the coalfields) constitute around 1 percent of all compensation claims,[29] the extremely low reported rates of occupational disease (in the few states which publish statistics) are largely a function of the low probability of receiving compensation and the widespread ignorance, both medical and popular, about occupational disease.[30]

The 1972 *President's Report on Occupational Safety and Health* stated that "at least 390,000 new cases of disabling occupational disease" develop each year. This figure, apparently derived from a projection of California workers' compensation data,[31] is probably far below the true incidence rate, given the barriers erected against compensating occupational disease. Dr. Thomas F. Mancuso recently discussed what happens with even the most obvious cases of industrial dermatitis:

> most . . . cases rarely . . . get reported as occupational disease
> unless disability has become so bad that a compensation claim is

considered. . . . I recall at least one small plant [where I learned of] at least 20 clear cut cases of occupational dermatitis, none of which has filed a record or claim with the compensation agency. I have also seen similar cases in which the worker is paying all the medical bills and neither the company nor the compensation agency provided any medical coverage.[32]

The chances of recognizing and winning compensation for an occupational respiratory disease are very slim, according to Mancuso, since individual workers usually have different personal physicians untrained in job health hazards who rarely perceive the exposed workers' diseases as due to a common cause. The only health specialists who view the plant as a unit, the plant nurse and physician (if such exist), are responsible to the company. If a worker comes to the company dispensary with a respiratory complaint, typically,

a report would only be made if a compensation claim was filed and if company liability is established. But the compensation liability for lung disease is extensive and "if this case is allowed, then other workers will also want to file a claim." The doctor is placed in the difficult position of wondering how long his arrangement with the company will continue.[33]

Efforts to make occupational diseases more readily compensable have almost always been crushed by business and insurance interests in the state legislatures, as, for example, when silicosis became a national scandal in the 1930s, or when Wisconsin boilermakers tried to make occupational hearing loss easily compensable.[34]

The Death Toll from Occupational Disease

Attempting to estimate the occupational disease death rate based on workers' compensation reports is like trying to measure the Mississippi with a bucket as it flows past. Projections from California compensation data on occupational disease (supposed to be the most reliable of their kind) would attribute only 1,200 to 6,000 annual deaths nationally to occupational disease, depending on whether or not heart disease and strokes are included in the definition.[35] But where independent confirmation is available, as in the case of asbestos-related disease, the California total is many times lower than the true figure.[36] Such occupational diseases as silicosis (a degenerative lung disease caused by fine sand), brown lung (a cotton dust disease), hearing loss, and cancer have been effectively incompensable because employers and insurance companies have controlled the compensation criteria.[37] Only the coalminers, subject to black lung, have been able to pierce the compensation barrier significantly.[38] Victoria Trasko's careful study, "Silicosis, A Continuing Problem " (1958), concluded that "despite all the interest, research, and activity since the 1930s, the precise prevalence of dust diseases remains a mystery."[39]

Since neither the compensation system nor physician or employer reporting can be relied upon to provide accurate estimates of the national scope of the occupational disease problem, methods independent of those sources must be encouraged. In 1972 the National Institute for Occupational Safety and Health (NIOSH), in the *President's Report on Occupational Safety and Health*, estimated that "there may be as many as 100,000 deaths per year from occupation-caused diseases." This estimate was based on the excess over the expected number of "nonviolent" deaths observed in several groups of workers, presumed to be caused by factors related to work.[40] Clearly, the estimate in the *President's Report* is inexact, but it implies that the number of deaths reported by the workers' compensation systems is understated by a factor of ten. It is certainly plausible that 100,000 deaths are caused annually by job-related diseases, if heart disease, lung disease, and cancers of the lung, kidney, liver, and bladder are even partially linked to occupational exposure. A federal study, cited by Health, Education, and Welfare Secretary Joseph Califano, estimated that from 20 to 40 percent of cancer deaths were caused by on-the-job exposure.[41] Similarly, a medical and industrial hygiene survey of workers employed in a sample of workplaces with under

150 employees in the Pacific Northwest found that a very high proportion of the workers suffered from "probable occupational disease," defined as "manifestations of disease ... consistent with those known to result from excessive exposure to a given injurious agent; this injurious agent is present in the patient's working environment and significant contact in course of usual duties is likely." The authors of this Seattle study comment that "in fact, the incidence rate for California production workers in manufacturing industries similar to those of [this] pilot study was about 15 per 1,000 workers [1.5 percent per year] as compared to the prevalence rate of 284 per 1,000 workers [28.4 percent]."[42] Of the 451 cases of "probable occupational disease" found by this study, 399 were identified only by the medical survey, recorded neither in the employer's log (as required by OSHA regulations) nor in compensation records. The study concluded that there is a "vast reservoir of unreported occupational disease" in the working population.[43]

Counting the Casualties: The Case of Missouri Before OSHA

Missouri's industrial casualty reporting system before OSHA exemplifies the pitfalls of depending on the workers' compensation system as a source of reliable data. It is likely that more than half the deaths caused by job accidents and almost all the deaths from occupational disease go uncompensated in Missouri, or, in other words, up to 95 percent of the deaths caused by working conditions in Missouri never reach the compensation system. For example, only 115 "first-injury reports" of death (supposed to be filed by employers after a compensable injury or death) were filed in 1968.[44] Yet, projections based on the assumption of 15,000 to 115,000 job-related deaths nationally, suggest that 380 to 2,780 people were killed by work-related conditions in Missouri in 1968. This assumption, admittedly hypothetical, is based on the National Safety Council's estimate of around

15,000 deaths caused by work accidents and the NIOSH estimate of 100,000 deaths caused by occupational disease. In 1968 Missouri's workers comprised 2.4 percent of all U.S. workers.

Estimating the number of injuries on a state-wide basis from workers' compensation statistics would not give figures comparable to those from other states because definitions of injury and rates of reporting vary so much between states. For example, the number of first injury reports per hundred workers per year ranges from two (Michigan) to thirty-seven (Alaska). Missouri, which requires reporting of "only injuries requiring medical treatment beyond first aid" showed a rate of nine "first reports of work injuries per hundred covered workers" in 1965,[45] perhaps one-third or one-half of what might be expected by extrapolation of the Gordon Report's "serious injury" criterion (see p. 42).

Occupational diseases were supposed to be reported by doctors to the Division of Health of the Missouri Department of Public Health and Welfare, but the law was not enforced. Mr. L. F. Garber, Director of the Section of Environmental Health of the Division of Health, wrote that due to the "infrequent receipt of a report of occupational disease" no statistical summary of such reports was considered necessary. He went on to state, "We have found the reports of injuries filed with the Division of Workmen's Compensation to be a reliable source of occupational disease statistics. Virtually all of the reports involve dermatitis and include only those cases severe enough to justify a claim for compensation."[46] Only 1,003 cases of occupational disease were reported in Missouri for 1968 through the compensation system. However, extrapolating the results from California compensation data or from the Seattle study, and assuming that only one-fourth of the cases of "probable occupational disease" identified in the latter develop each year, results in an expectation of anywhere from 4,000 to over 120,000 cases (based on an estimated 1,800,000 workers in Missouri in the late 1960s) involving much more than "dermatitis." It is thus likely that 75 to 99 percent of occupational disease cases in Missouri never reach the compensation system.

Alternatives to Corporate Control over the
Official Body Count

One way to deny a problem is to keep no records about it. Another way is to design a record-keeping system which shows that the problem is being resolved, whether or not it actually is. Official government statistics are important because they are the figures everyone hears about a problem, and thereby serve to define the attention which should be devoted to its resolution. The same holds true for the National Safety Council's statistics, which maintained a quasi-official status for decades. A "problem" responsible for 15,000 deaths a year is a great deal less important than one which kills 115,000. It is natural for corporate-dominated groups to direct attention to the smaller figure,[47] however meager its scientific merits. The problem of occupational disease acquired the dignity of a federal estimate of its national incidence and severity only in 1972, sixty-four years after the first federal estimate of the number of workers killed on the job. However, even the Bureau of Labor Statistics has little confidence in its estimates of occupational disease incidence.[48]

A number of different approaches have been used to get more accurate information about total annual casualties nationally, but as long as they are collected by employers they will constitute a cover-up of the true incidence. The effort made by labor lobbyists to broaden the definition of a reportable injury or disease under OSHA regulations, although successful, missed the point, for the forms are still filled out by management. Every independent survey has shown that management cannot and will not register all "reportable" injuries. Even by National Safety Council standards the OSHA-collected figures massively understate the number of accidental deaths. Part 1904 of its regulations requires that all work-related deaths be reported to OSHA within forty-eight hours. In 1972 only 3,000 deaths were reported by this route, so few that OSHA itself rejected the figure. Thus there must be comparable underreporting of the job injury total reported by the same system.[49] Moreover, the Bureau of

Labor Statistics' estimates for occupational disease incidence are very low; for example, OSHA reported that 3 percent of the cases reported by employers in 1974 under the
employer's log system were occupational illnesses, for an
incidence rate of 0.4 per 100 full-time workers.[50] Whatever
usefulness the employer's log has could be met, with a
welcome cut in paperwork for employers, by requiring only
the supplementary record of workers' compensation claims,
which has to be maintained for insurance purposes in any
case. The supplementary record should be opened to inspection by workers at any time. In its effect the OSHA employer's log system has been a more pernicious hoax than
even the NSC's Z16.1 method: it gives government sanction
to what is essentially a privately administered casualty reporting system.

What should replace it depends on the purposes and costs
of the record keeping. Sample surveys of households could be
checked against industry reports to arrive at OSHA estimates
of injury incidence. The Current Population Survey of the
Bureau of the Census already collects monthly data on employment and unemployment and other subjects. The National Health Survey has asked a series of questions on job
injury and disease which could be adjusted to fit the OSHA
statistical reporting requirements. Its questionnaire could be
adapted to OSHA injury recording standards, taking advantage of an already existing survey at a minimal cost.[51]

Another system which could be used to report and combat
industrial injuries and acute occupational diseases is the
National Electronic Injury Surveillance System (NEISS), part
of the Consumer Product Safety Commission's effort to
reduce accidents involving consumer products. NEISS functions as follows: at a representative sample of over 100 hospital emergency rooms throughout the United States, all
accident victims (or whoever brought them in) are queried as
to whether or not the accident involved a consumer product,
and what that product was. These reports are coded and
transmitted daily to a Washington computer where sum-
maries are prepared automatically. On this basis, research
priorities are decided upon for the field investigators, based

on frequency and severity data for the injuries. The field investigators try to be on the scene of an accident within seventy-two hours.[52] Despite the fact that many serious industrial accident victims are treated in private industrial clinics or company medical departments, the fact that many appear in hospital emergency rooms should make this an adequate basis for additional data.

Extending the NEISS program to industrial accidents by assigning some OSHA compliance officers to carry out accident investigations using NEISS methods would provide a data base completely independent of management. Since it would use an already existing system it would not be expensive to set up, and it would lead, as it has in the consumer field, to an abundance of practical and publicly available information on the redesign of industrial equipment.[53]

Methods such as those developed in the Seattle study, combining medical examinations of workers with industrial hygiene surveys, appear to be the only way to develop accurate measures of the incidence of occupational disease in the workforce. Unfortunately, however, proposals to carry out such a study on a national sample have been dropped by NIOSH.[54]

Estimating deaths due to occupational disease is usually carried out by epidemiologists. Most of the occupational diseases that kill Americans do so after a long-term, low-level exposure to on-the-job pollution. Emphysema and lung cancer, for example, have multiple causes, so that doctors usually do not attribute them to work exposure. Given the low priority of occupational health in medical training, it is difficult to excite interest in the prevention of work-related disease. Occupational diseases are rarely mentioned on death certificates, so the fact that a particular factory or trade is hazardous becomes clear only when the death rates of those exposed are analyzed, or when a detailed study is made of many workers and their work environments.

Historically, employers and insurance companies have been very jealous of their medical and personnel records, particularly when outsiders are on the trail of an occupational disease.[55] Companies often use these records to weed sick workers out of the workforce if they fear such workers might later file an occupational disease claim against them. One of the chief issues of the 1974 strike against Shell Oil Company by the Oil, Chemical, and Atomic Workers Union was over access to medical and personnel records which could reveal sickness and mortality profiles. Despite the strike, very little information has been released to the union.

In his asbestos research, Dr. Irving Selikoff and his associates were able to break the companies' monopoly of medical information by working with a craft union which had its own welfare and retirement system. But only a very small proportion of the U.S. workforce belongs to unions with such records. Dr. Thomas F. Mancuso's solution to the problem of finding out when and where people work and die has been to examine the records of the universal retirement fund: the Social Security Administration. Through the use of Social Security's quarterly reports it is possible to determine the name, Social Security number, birthdate, sex, place of work, and date of death for any employee since 1937. Thus "the conduct of epidemiological studies of any plant covered by Social Security since 1937 can be carried out by Social Security internally by its own research staff [overcoming] the destruction of the personnel records during the past 36 years."[56] Mancuso recommends that the Social Security research department be empowered to report the mortality rates of industries by type and geographical area as a regular part of its operations every three years, thereby providing a basis for "more refined" approaches. His technique has its virtues: it would not impose new record-keeping requirements on employers which they would have every incentive to ignore or falsify; and it would use existing data of high quality which is outside of corporate control. Use of the Social Security records would help identify dangerous industrial operations and industries automatically by a computerized mortality audit. Once the system was set up, it would for the first time be possible to generate a relatively accurate estimate of the excess deaths caused by working conditions in the country as a whole.

The implementation of honest national data-gathering

procedures is imperative in order to avoid wasting limited health and safety resources on trivia. The struggle for an accurate official casualty count should become an important part of the new health and safety movement, for the experience of seven decades has proven the ability of the compensation-safety apparatus to lie with statistics.

|3|

How Cheap Is a Life?

A Puerto Rican life is worth from $11.34 to $28.86 a week to the compensation authorities. A worker killed on the job in Mississippi will cost the boss's insurance company no more than a flat $15,000. A longshoreman's widow will receive from $79.60 to $318.38 per week as long as she lives. In Illinois, the family of a worker killed on the job may collect up to $205.00 a week. In North Dakota they add $7 a week to raise a child.[1] Business complaints about the high cost of workers' compensation for death and disability conceal the fact that it replaces less than 10 percent of the income foregone by the victims (see Table 5, Appendix 1).[2] Along with the pain, workers and their families have borne most of the financial burden; the public picks up much of the rest through taxation to pay for welfare and social security for those no longer able to work.

Compensation has always overshadowed prevention. The sharp increase in total expenditures on occupational health and safety since 1970 has done little to change this. In the preventive sphere, programs such as industrial inspection still channel most of their efforts toward safety rather than health issues, although medical research is an exception. Much of the blame for the underemphasis on health must be laid on the system of workers' compensation, which still makes it nearly impossible to collect medical and disability benefits for job-related diseases. Although the Occupational Safety and Health Act of 1970, discussed in many contexts throughout this book, is the most important health and safety law ever passed, and although in conjunction with other tactics it has sometimes led to great improvements in working conditions, its passage has not changed the fact that 85 percent of the expenditures in health and safety are made by the private sector, which dominates public policy (see Tables 1–3, Appendix 1). It was business lobbying, for example, that finally convinced Congress and President Ford to amend OSHA to exempt farms with under ten employees from inspections, and to forbid the fining of employers found to have less than ten "nonserious" violations of health and safety regulations.[3]

Like wages, the issue of job health and safety is a class issue. Clean working conditions cost money. Transferring the costs of hazardous conditions from the working class to the capitalists by enforcing preventive measures or by driving up compensation payments cuts into profits unless those costs can be successfully resisted or foisted onto the public. The multibillion-dollar black lung benefits program for coal miners, for example, has been paid for mostly from the federal treasury.[4]

Large American corporations, typically unionized, have shown less resistance to federal regulation of working conditions than small businesses because they can influence enforcement and rule-making in Washington. Usually self-insured, these corporations can bypass high-cost workers' compensation insurance premiums. When union or public pressure make it impossible to postpone improvements in working conditions, big corporations can raise prices to pay for them, thus transferring the cost to the public. At the same time, corporate giants can tap insurance reserves (accumulated from premiums of small- and medium-sized businesses) through the stock exchange and other capital markets, sources of credit normally unavailable to small enterprises. Smaller, more competitive businesses, operating closer to bankruptcy, usually have to pay for nonproductive safety equipment and compensation insurance increases directly out of profits. Since stringent enforcement of the

OSHA law for small businesses would drive a number of them out of business, they have opposed the law more vehemently than large corporations.[5] Despite differences in emphasis, however, business as a whole usually unites to oppose the implementation and enforcement of stricter workplace standards and significantly improved compensation benefits.

Accident-Prevention Programs: State, Federal, and Private

Before OSHA, the efforts in most states to protect workers through industrial inspection services were minimal. In the middle 1950s it was estimated that state and local occupational health programs together spent under $3 million per year.[6] In 1969 Secretary of Labor Willard W. Wirtz testified that total state safety department expenditures were about $23 million in 1965—about 33 cents per worker nationwide, ranging from 2 cents in Texas and Oklahoma to $2.11 in Oregon.[7]

Usually industrial inspection was practically a dead letter. Evidence from hearings and other sources on Ohio, Missouri, New Jersey, Pennsylvania, Indiana, and Massachusetts[8] shows that state programs were understaffed, their personnel was badly trained, and they were reluctant to carry out enforcement proceedings. Moreover, inspectors in such programs usually worked in collusion with management and gave advance notice of important inspections.

In a 1970 study of the Division of Industrial Inspection in the state of Missouri, I found that appointments of inspectors were totally dependent upon patronage politics: between 1925 and 1969, all division employees were replaced each time the governorship changed parties. I concluded at that time that only special pleading by workers or unions would draw attention to the bad conditions in any specific plant. Even after inspection and writing of a work order to correct the fault, nothing was done by way of enforcement, although the legal weapons were there. No one in the division of Industrial Inspection, including George Flexsenhar, the director, could remember any fines or the sealing of any machinery under the provision of the law. According to Flexsenhar, enforcement would have to take place through the Attorney General's office, a complex procedure that would divert resources from more pressing needs. When asked what would happen if he shut down a dangerous department of the state's biggest lead smelter, Flexsenhar replied that "the governor would be on the phone this very afternoon" to make him retract the order.[9]

Even states with relatively substantial programs before the passage of OSHA drew criticism from workers. Daniel W. Hannan, president of the United Steelworkers' local at the "largest byproduct coke plant in the world," was angry at the Pennsylvania Department of Occupational Health for refusing to give his union access to information about conditions in his shop. The department's survey had shown that workers in the Clairton coke ovens were exposed to airborne cancer-causing coal tar pitch volatiles at up to fifty-two times the "threshold limit value" recommended for eight hours of exposure by the American Conference of Governmental Industrial Hygienists. When he tried to get an official copy of the report of these results for his union, Hannan was given a labyrinthine runaround:

At first this department denied the existence of a report or that an investigation had taken place, but anyway, I was told to go to the company and request a copy of the report from them. This I refused to do. I felt my request was a reasonable one of the health department and that their actions were very strange in light of the fact that the Department of Occupational Health is a tax-supported agency of the State government. I then contacted my State senator and the director of the Steel Workers legislative committee for further action on my request. Our State legislative director found all doors of cooperation were closed with the State government and finally the director contacted the minority leader of the State senate and the speaker of the house. The speaker ... was also the chairman of the State appropriations committee and he and the minority leader of the State senate requested a copy of the report and they were refused. The chairman of the appropriations committee threatened to withhold

funds slated to operate the health department for the new fiscal
year. The copy of the results of the survey were then released and
I was sent an original copy.[10]

If Hannan had not been politically sophisticated, persistent,
aware of the health issue at stake, and president of a large
local of a powerful union he would never have seen a copy of
the report. An ordinary worker would not have had a chance.
Only now, after a decade of struggle, has the steel industry
promised to clean up working conditions in the ovens over a
period of years.

On the federal level, standards-setting, inspection, and
enforcement programs were almost totally inactive before the
OSHA law. Under the Walsh-Healey law of 1936, private
employers with over $10,000 in federal contracts were re-
quired to maintain federally mandated health and safety
standards. But 1969 Senate hearings revealed that those
standards were not promulgated for over 25 years.[11] In fiscal
1969 only 5 percent of the 75,000 firms covered by Walsh-
Healey were inspected by the Bureau of Labor Standards. Of
33,378 violations discovered, only thirty-four formal complaints were lodged and only two firms suffered the ultimate
sanction of blacklisting from future federal contracts.[12]

In the private sector, preventive efforts by insurance
companies have generally been ineffectual. Although the
insurance industry claimed to have spent $26.3 million on
inspections in 1966, for example, at the 1968 Senate hearings
the general manager of the National Council on Compensation Insurance (NCCI) was unable to specify where the
money went. The few insurance company inspections actually carried out are concentrated in large factories. The
former practice of schedule rating, by which premiums were
partially determined by the quality of safety equipment and
practices, was abandoned because of employer resistance to
what they saw as "policemen in the plants."[13]

As for private employers, we have no estimates of the
amounts spent on ventilation, machine guarding, training,
and other measures to protect workers before OSHA. Even
so, it appears that preventive expenditures were always much

less important than what was spent on compensation. The passage of the OSHA act led to a large initial jump in total expenditures, although such expenditures seem to have leveled off as a proportion of Gross National Product since 1972. Figures gathered since 1970 show that the private sector continues to dominate employment and spending in occupational safety and health (see Tables 1–3, Appendix 1).

Keeping Up the (Compensation) Premiums

Examination of the financial structure of the workers' compensation system is crucial to understanding why many people get hurt, how much compensation they get, and whether their medical expenses are paid. Two-thirds of the money earmarked by public and private programs to deal with job health and safety flows through the compensation system (see Tables 1–3, Appendix 1).

Workers' compensation is usually carried by property-liability insurance companies. These companies, which obtain an eighth of their premium income from workers' compensation programs, earned $45 billion in net premiums and controlled assets of $78 billion in 1974. These assets (equal to 5 percent of U.S. finance capital)[14] were invested in all sectors of the economy, including $18 billion in state, special-revenue, and tax-free municipal bonds, and $17 billion in common stocks.[15] Linked by myriad formal and informal ties to large banks and corporations,[16] the insurance industry thus wields great power over compensation regulations enacted by state legislatures. Cooptation of government officials to the insurance point of view occurs through campaign contributions and guarantees of high-paying jobs in insurance, and, where necessary, through bribery, refusal to sell money-losing lines of insurance, and even manipulation of state governments' ability to borrow money.[17]

The enormous power of the insurance companies in workers' compensation legislation developed through a series of successful struggles to establish and expand their roles, as well as stabilize a position of privilege, in compensation programs. When compensation programs were first set up, there was a question in many states whether private insurance companies would be permitted to handle them. Coalitions of unions, socialists, and such groups as the Non-Partisan League of North Dakota opposed such a role, and between 1911 and 1915, thirteen states established state-owned compensation insurance funds. But despite repeated efforts in many states, only five more state funds were set up, most of them in competition with private carriers.[18]

Profits as well as business had to be secured, however, and these were at first threatened by rate wars between competing companies. An early advisory organization proved unable to control competition, so an association of state insurance commissioners called the National Convention appointed a committee to confer with insurance companies on stronger measures to prevent "chaos" in rate making. The result was the formation in 1923 of the National Council on Compensation Insurance, which in most states to this day functions as the agency responsible by law for calculating "reasonable and adequate" premium rates, with the full cooperation of state insurance commissioners.[19] These openly monopolistic rate-setting practices came under federal antitrust attack in the 1940s, when the U.S. Supreme Court handed down its Southeastern Underwriters decision holding that insurance was "commerce" and thus subject to the federal antitrust laws. The insurance industry viewed this decision as a direct challenge to "the right of (rate) bureaus to make and administer rates," and convinced Congress to pass the McCarran-Ferguson Act exempting the insurance industry from antitrust laws "where some semblance of state regulation existed." In states lacking such legislation, the insurance industry created it, making itself completely immune from federal antitrust prosecution.[20]

Since employers, after some initial reluctance, left rate setting to the insurance industry,[21] it has in effect become completely self-regulating. The National Council on Compensation Insurance now sets compensation premiums based on benefit levels set by state legislatures in most states, and states with independent bureaus, such as California, cooperate closely with this organization.[22] NCCI operating expenses of $8 million per year come from levies on the compensation carriers.[23] All accounts of the relationship between the insurance industry and government regulators show complete control of state insurance commissions by the insurance business: the well-known failure of Insurance Commissioner Denenberg of Pennsylvania to effect any lasting changes favorable to consumers is one example of the enormous political power of the insurance companies.[24]

Keeping Down (Compensation) Benefits

Although millions of workers have benefited from workers' compensation programs, particularly those with medical expenses and acute, temporary disabilities, the system overall represents a tremendous success for employers in transferring the costs of occupational casualties onto workers, their families, and the public. In general, those who are hurt the most have the smallest percentage of lost income replaced by compensation. Workers who suffer permanent disability from back, head, and multiple injuries for which benefits can vary have only a small part of their income losses replaced by workers' compensation; these are "nonschedule" injuries, meaning there is no specific amount of money to be paid, and the worse the disability, the smaller the proportion. In many states, death benefits for workers' survivors total only two to four times the median annual wage.[25]

Although benefit levels in 1974 finally surpassed pre-World War II levels as a percentage of covered payroll (see Table 4, Appendix 1), private insurance companies continue to sell most of the coverage, and their expenses and profits still absorb between 40 and 50 percent of the cash flow from premiums. A glance at the membership and conclusions of the *Report of the National Commission on State Workmen's Compensation Laws* (1972) suggests that change so far has been controlled by those whose interest lies in maintaining the status quo, a conclusion reinforced by the conservative nature of recent reforms (mostly catch-up increases in benefit levels and some broadening of statutory eligibility). Not even the modest reforms recommended by this report had been accomplished in most states by 1976.[26] Employer and insurance groups have resisted efforts to make common occupational diseases easily compensable, and rehabilitation of severely injured workers to the point where they recover their former earning power and bodily integrity is rare. To date, the very notion of setting federal performance standards for state workers' compensation systems has been rejected.[27]

Despite recent increases in benefit levels, estimates based on data collected by the National Commission on State Workmen's Compensation Laws, by private insurance sources, and by recent research shows that less than a quarter of the total income loss *of those entering the compensation system* is replaced by workers' compensation benefits. Four-fifths of those entering the system (mostly those disabled for less than eight working days), receive only medical benefits. Those receiving compensation recover an average of slightly more than one-fifth of their income losses from job-related disability (see Table 5, Appendix 1). If all the lost income of those injured workers entering the compensation system had been replaced, the amount paid out nationally would have totaled $18.2 billion, rather than the $4 billion actually paid out.[28] Yet even the $18.2 billion figure is much smaller than the true wage losses caused by industrial injury and disease, since so many permanent disability and illness cases are effectively excluded from compensation.

Not surprisingly in an economy with a permanently high unemployment rate, little effort, public or private, is made to rehabilitate disabled workers, since they are easily replaceable. A study of 121 permanently and very severely disabled men in the New York City area who had been injured at least four and a half years previously found that they and their households had "suffered severe economic attrition and adverse changes in quality of life." Only a third had experienced any contact with the state rehabilitation agency. Of the one in eight who had learned a new trade, only 30 percent did so at a state rehabilitation center, and less than 10 percent could return to their former trade.[29] Other studies generally agree that permanently disabled workers are simply ignored by the rehabilitation agencies and often by their former employers.[30]

The emotional strain on the seriously injured is considerable. At the time when they are unable to work and often in great pain, their income is cut off and they are forced to try to deal with rapacious attorneys, insurance agents, and company doctors. Many think of suicide, and some try it. Mary Quartiano, a waitress in San Diego with a permanently disabling back injury, who has now begun to help others through the compensation labyrinth, writes:

> Unfortunately, most of the people who have come to me so far have been messed over by their employers, their unions, their attorneys, and many physicians, the same way I was. This takes many hours of counseling and is an extreme drain on my own emotions, which at times are not on an even keel. This is understandable when you realize that sometimes the same professionals messed them over that messed me over.
>
> Recently I have been helping a young black man who has had many problems with the Comp. system. When he first came to me I wasn't sure I could help him because he was absolutely hysterical. I now have him calmed down, in the hands of a good doctor, and will send him to a good attorney when the time is right. Right now we are just buying time until he can be put into a rehabilitation program. Timing is everything at this stage of the game.
>
> It seems that I have to be everything to everyone; psychiatrist, doctor, and attorney, none of which I am. And sometimes it is a little overwhelming.[31]

In the absence of independent and skilled analyses of insurance company finances, it is hard to know how much money the industry really makes. A conversation I had with representatives of Accountants for the Public Interest in 1977 revealed that insurance industry accounting is a mystery even to accountants. Despite property-liability companies' claims of underwriting losses in eight of the past ten years (with 1974 and 1975 the worst years), their reported assets doubled, from $41 billion to $82 billion between 1965 and 1974. In 1977 profits soared.[32] Whatever efficiency for insurance companies is supposed to mean, it certainly isn't efficiency in returning benefits to compensation claimants. Nationwide, the benefit/premium ratio for private compensation carriers in 1974 was 0.53, compared to 0.72 for state funds.[33] In addition to what private insurance carriers retain as profits and expenses, about one-third of the incurred losses nationally are due to medical and hospital expenses, so that only 40 percent of the funds which flow into the private workers' compensation system are paid directly to the victims or their survivors. From this amount lawyers and sometimes doctors must be paid in order to contest a claim. My own investigations confirmed that the lawyer usually takes from 10 to 40 percent of the settlement.[34]

In Ohio, Washington, West Virginia, North Dakota, and Wyoming, where private compensation insurance companies are not allowed to operate, the state funds have managed a benefit/premium ratio of 0.94 over the years, with Ohio the leader at 0.96. Although it is claimed that state fund benefit/premium ratios are lower because they deliver fewer services and because they are exempt from taxes, clearly the fact that they have no selling expenses and do not make a profit has something to do with their apparent efficiency in getting benefits to workers. Workers' compensation benefits could be nearly doubled at no new cost to employers by eliminating the role of the private insurance companies. A good way to start would be to investigate why it costs Ohio just over 4 cents—at 4 cents for every 96 cents returned—to return a dollar of benefits to work accident victims, while it costs the private insurance companies 89 cents—at 47 cents for every 53 cents returned—to return a dollar of benefits.

Insurance companies and employers are able to keep benefits low because of their effective lobbying apparatus in state legislatures which set benefit rates[35] as well as compensation procedures. Since most states require waiting periods of three to seven days before compensation is provided,[36] four-fifths of the workers entering the system draw no benefits beyond the payment of their medical bills (see Table 5, Appendix 1). Seriously injured workers applying for major permanent disability benefits are up against tremendous odds. They usually have very little information about their rights to compensation, even if they are unionized, and they are further victimized by lawyers and physicians.[37] Insurance company manuals which recommend prompt payments for scheduled injuries such as chopped-off fingers or hands give little credence to "subjective complaints (which) cannot be proved or measured" such as headaches and back pains,[38] even though such injuries may be much more disabling. In many back and head injury cases where the insurance company contests the worker's claim, the terms *traumatic neurosis*, *psychogenic neurosis*, and *neuropsychiatric ailment* often crop up in the medical reports. These expressions have come to replace the term *malingerer* in insurance and corporate medical jargon, and are part of the same attempt to shift blame from the employer to the worker. In fact, according to a study by Richard E. Ginnold, insurance company physicians hardly ever diagnose traumatic neurosis among hand and eye injury victims because it is impossible to deny the existence of such injuries and their connection with work accidents. He believes very strongly that there is little faking of back and head complaints.

The cases for which "traumatic neurosis" is implied are back injuries (and to some extent head and multiple injuries). Yet if we examine these cases, most were high earning individuals before the injury, and were in occupations requiring hard work—construction, truck driving, laboring. They would not predictably be candidates for neurosis prior to the injury. Most have never had another workers' compensation claim. It is also hard to believe that all the potential neurotics would be lumped into one injury group.[39]

Mary Quartiano provides confirmation from her own experience: "They almost convinced me I was paranoid or neurotic for pressing a claim. I think that's how they try to wear people down."[40] Bob Fowler, formerly a carpenter at the giant United Airlines repair facility at San Francisco International Airport, and a member of Machinist's Lodge 1781, was tailed by insurance investigators at his home after he hurt his back at work. He believes that they were trying to photograph him doing work he could not do on the job.[41]

Many workers find employers reluctant to hire them or take them back after head and back injuries, for fear of another compensation claim. For some, a back injury prevents them from making love or even from sleeping in a bed.

Women suffer "far greater losses than males for all injury types," according to Ginnold's study, and they are brutally victimized in all aspects of their lives by disfigurements which would not repulse people as much in a man.[42] One woman who had lost some fingers was told not to come to church without gloves, and another caused a department store clerk to vomit at the sight of her mangled hand. Even when these women are physically capable of working, they are simply shut out when potential employers see the injuries. One woman who lost three fingers from her left hand and crushed her right in a punch press believes her husband left her because of the injury. In an interview she stated:

> The boss came to see me in the hospital and had me sign settlement papers for $8000. I feel cheated. I had no lawyer. The employer never offered me work after the accident. I looked for a job for one and a half years, then quit. I can do anything except type, but all prospective employers give me the runaround. The Department of Vocational Rehabilitation . . . sent me to workshops for retarded people, but I'm not retarded.[43]

From its founding by large corporate interests, the compensation system has carried out the interests of employers and insurance carriers in keeping down benefits. Restrictions began to be made in the scope of injuries covered, and old legal defenses supposedly discarded by the advent of workers' compensation made their appearance under new guises. The "assumption of risk" doctrine crept back in as compensation courts "tended to exclude cases where the stress was normal to the job." The defense of "contributory negligence" was in effect reinstated by calling it "willful and intentional misconduct."[44] In Missouri, for example, benefit levels could be raised or lowered by 15 percent if it could be shown that an employer or employee had not followed safety rules. This regulation almost always worked in favor of the employer, for "years of rule refinement, growing out of hundreds of accidents and many lawsuits, have made infinitesimal the number of mishaps that a higher manager could not show as 'due to not following the rules' . . ."[45]

Small claims are paid off quickly, while large claims are delayed or denied. The medical-only cases, which account for 80 percent of those entering the system, are paid promptly to the physicians providing the care. This avoids harassing millions of workers with medical bills; and the physicians, often chosen by the insurance carrier or employer, tend to become favorably disposed toward the carriers in the advent of litigation. Besides, the amounts involved, averaging $45, are too small to justify haggling (see Table 5, Appendix 1). Major and total disability and death cases, by contrast, average $20,000 each. Though comprising under 1 percent of all cases entering the system, they account for 39 percent of the benefits paid (Table 5). As a consultant to the National Commission on State Workmen's Compensation Laws pointed out, "All forces tend in the direction of underpaying the seriously injured worker. . . . The worker with an injury which keeps him out of work for a long time and who is . . . in most need of full benefits is more likely to be substantially underpaid—by his own 'choice'—because of the financial pressures on him."[46] Again, the more you need the money, the less likely you are to get it.

Occupational Diseases: The Cases that Get Away

Occupational disease compensation has always been more difficult to obtain than injury compensation for a series of reasons, all related to the size and uncertainty of costs which would be transferred to employers and insurers. The latter are also afraid that if benefits got too high or too easy to collect, it would be harder to keep people at work and to maintain discipline in the shop.[47] The U.S. Chamber of Commerce estimated that the liberal provisions for occupational disease compensation proposed by Senators Jacob Javits and Harrison A. Williams in 1973 might raise annual compensation costs seven times, to $40 billion a year,[48] a figure that makes sense if occupational disease kills 100,000 workers a year.

Evidence that occupational disease is ignored by the compensation system comes from a National Commission study using data from a Social Security survey on work-limiting disability. This study examined a sample (representing 2.2 million men nationwide) who had been disabled for at least three months by a work-connected disability. It was found that only 24.7 percent of these men had ever received any income from workers' compensation, and only 8.7 percent were still being compensated at the time of the survey. To the commission author, the "outstanding puzzle" in the data was that 68.5 percent of the men had never even bothered to apply for workers' compensation, though they considered themselves victims of a job-connected disability. Of these nonapplicants, 41 percent were suffering from lung disorders and 40 percent from disabling bone or joint conditions of some kind.[49] That so many men suffering from work-related disabilities neglect to file for workers' compensation suggests a gigantic reservoir of uncompensated injuries and diseases.

Occupational diseases and the hazards that create them are often "invisible" for the following reasons:

1. Powerless workers who complain about work poisons are ignored or fired, and they are purposely kept ignorant of the hazards of their jobs.[50]
2. A work-related cause-and-effect relationship often is not obvious to the worker or family physician, particularly in considering diseases caused from long-term, low-level ex-

posure to work poisons, such as many cancer-causing agents. Often proof of a work relation requires long, costly studies of many workers.

3. Many workers are exposed to multiple hazards about which little is known, or about which information has been systematically forgotten or misrepresented by the medical profession and researchers in the pay of corporate interests. Business and insurance lobbyists in state capitals struggle to keep legal definitions of occupational disease narrow. Particular interests, such as the coal industry, the Rohm and Haas Company, and the asbestos industry, have fought to make it difficult to get compensation for diseases characteristic of their production activities.[51]

4. The proliferation of tens of thousands of new chemical and physical hazards in the work environment (as well as in the environment at large) has run far ahead of research on their toxic effects. To date, very little is known about what such agents do to people: mammal testing of the acute and chronic effects of each chemical can cost from a quarter to a half million dollars, takes years, and is not always a reliable guide to human effects. Although a quick test for mutability costing a few hundred dollars already exists and is thought to be relevant in predicting the ability to cause cancer, mammal tests will still be necessary to check its findings.[52] Perhaps inroads into these problems will be made by the new Toxic Substances Control Act, passed in October 1976 after much watering down due to industry pressure.[53]

The effect of the huge new interest in occupational disease on compensation practices has been minimal. Since the passage of OSHA occupational disease claims have risen from "well under 1 percent" of all compensation benefits to about 1.35 percent, according to Andrew Kalmykow of the American Insurance Association.[54] For the average employer this amounts to one dollar in ten thousand of payroll costs.

Compensation's Irrelevance to Improved Working Conditions

Defenders of workers' compensation laws claim that "safety pays" for the individual company; that is, the present system furnishes automatic incentives to constantly improve conditions at the workplace in order to reduce the costs of casualties. The argument has two facets: one rationale holds that premium rate-setting practices for workers' compensation forces companies to try to minimize occupational disability; another maintains that the "indirect" costs of occupational hazards to a company are so high (two to twenty times the cost of workers' compensation) that companies can ignore health and safety only at great financial peril.[55]

Compensation insurance premiums for particular workplaces are set by assigning to an establishment a basic rate depending on its size and type of work. In California the rates per $100 of payroll vary from $.125 (auditing and accounting offices) to $39.70 (wrecking or demolition of structures) among over 800 rating categories, with most clustering around $1.00 or $2.00.[56] This industrial classification rate is then adjusted according to "whether injury loss control has been average, better than average, or worse than average . . . based on three years of loss experience."[57] Very small workplaces are excluded from this process because of its cost, but about 80 percent of all compensation premiums are paid by large "experience rated" companies.[58] The dollars saved by an improved "experience rating" are supposed to spur companies to greater efforts at prevention of work-related casualties.

In fact, the "experience modification" has been used by insurance companies mostly as a rate-setting device. It has had no value to employers trying to evaluate disability costs and allocate their preventive efforts because it arrives too late to affect company decision-making and because it measures only the overall performance of the work establishment. Empirical studies quoted by the Labor Department showed "considerable skepticism" of the value of the experience modification as an "impetus to safety," a conclusion shared by studies done for the National Commission.[59] With workers' compensation costs usually averaging close to 1 percent of payroll and occupational disease costs almost completely discounted by compensation, most companies have no financial incentive to carry out preventive activities unless compensation premium and accident rates are extremely high. The pulpwood industry, for example, with premiums running at 26 percent of payroll, has been one of the few industries publicly enthusiastic about OSHA's help in reducing accidents.[60]

The argument that the indirect uninsured costs to employers of occupational casualties are high enough to force companies to prevent injury and disease also turns out to be largely untrue. These costs are difficult to measure accurately because cost-accounting systems (where they exist at all) are not designed to include them. An insurance company survey of a number of industry safety specialists showed that "none of the respondents . . . indicated that they were using available data for purposes of (1) costing, (2) budgeting, (3) accountability."[61]

Only a fifth of the income losses from occupational injuries are recouped, and only by *those who enter the compensation system* (see Table 5, Appendix 1). Since occupational diseases are a much greater source of disability than work accidents, and since it is almost impossible to get disease compensation, it makes sense to assume that over 90 percent of the personal income lost because of poor working conditions is borne by workers, their families, and the public. With compensation cheap and high unemployment chronic, there is little market incentive for employers to prevent the slow poisoning or instantaneous disablement of healthy workers. Today the workers' compensation system is largely irrelevant as an incentive to better working conditions.

Another Seventy Years of Stagnation?

Despite renewed scrutiny of the workers' compensation system, its basic structure has remained unquestioned and unchanged. The National Commission on State Workmen's

Compensation Laws was itself chosen "from people inside the system," according to the commission's business-school chairman, John Burton.[62] Although the commission concluded that workers' compensation was "inadequate and inequitable," it recommended that "the States' primary responsibility for the program should be preserved."[63] The bill introduced in 1973 by Senators Jacob Javits and Harrison A. Williams to set national performance standards for state programs died in Congress. More recent versions of the bill have been watered down by striking out provisions which would have given workers the right to choose their own doctors in compensation cases and make compensation easier to secure for occupational disease. Even so, none of the revised bills has passed. Management and insurance interests are frightened by the multibillion-dollar benefits for coalminers under the various black lung compensation laws and by recent losses in the property-liability business, and are strongly opposed to federal intervention.[64]

More recently, the top economic policymakers of the Carter administration have tried to eliminate OSHA's enforcement function, resuscitating the theory that workers' compensation could provide the incentive for businesses to prevent accidents. A memo to the president signed by Charles L. Schultze, chairman of the Council of Economic Advisors, Stuart Eizenstat, assistant to the president for domestic affairs and policy, and Bert Lance, former director of the Office of Management and Budget noted that

> ... serious consideration should be given to totally eliminating most safety regulations and replacing them with some form of economic incentives (for example, an improved workmen's compensation program, or economic penalties tied to the injury rate), thereby redirecting OSHA resources to regulating health problems and coverage of emergencies.[65]

This memo also reveals that the administration regards reform of OSHA as the first attempt at replacing regulatory agencies with market incentives. Naturally, this new look ignores the fact that the insurance carriers and businesses are mounting a nationwide attack on compensation benefits, particularly for permanent disability problems.[66] And the AFL-CIO Standing Committee on Health and Safety maintains that the Carter administration has made no attempts to revive congressional passage of federal standards for workers' compensation.[67]

Except for intermittent lobbying efforts on the national and state levels, unions, activists, and the occupationally disabled have not been able to generate much momentum on compensation reform.[68] There has been little critical examination of insurance company expenses and monopolistic rate-setting practices, and the debate has just begun to deal with access to compensation and rehabilitation for the permanently diseased and disabled. No public figure has openly considered the abolition of the private sector in workers' compensation and the integration of its tasks into Social Security and a national health plan. With living standards falling and social expenditures under heavy business attack, it is doubtful that workers will be helped by an administration committed to bolstering the private role in social welfare and cutting back the direct regulation of business.[69] In occupational health and safety, the old pattern of foundation and government studies which led to cosmetic reform is repeating itself after seventy years.

|4|
The Compensation-Safety
Apparatus

Despite the passage of OSHA, 85 percent of the funds for health and safety are still spent by private businesses and insurance carriers, which also have an important influence on the allocation of government monies in the area (see Tables 1–3, Appendix 1). As a result, almost all of the professional specialists in health and safety are directly employed by business. Most of the remainder usually work for federal or state governments, where they tend to be wary of criticizing employers. The constant struggle to minimize health and safety costs colors all the activities and attitudes of those employed in the private sector, despite frequent lip service to the ideals of prevention and adequate compensation.

A half-dozen organizations have for decades had an important role in defining the business response to the problems created by dangerous working conditions, but a description of the activities and financing of such groups is hardly adequate to characterize the scope of the business role in occupational safety and health. In fact, every insurance claims representative, every corporate industrial hygienist, safety engineer, and medical director functions as part of the compensation-safety apparatus. In order to understand the workings of this system it is necessary to look at a number of topics: the ideas which guide management safety practice; the activities of the National Council on Compensation Insurance in setting insurance rates; the changes in the business role as standards-setting moves from private to public arenas and from an emphasis on safety to an emphasis on health; and the training, attitudes, and practices in the principle occupational health specialties, illustrated by a glimpse inside the medical department of Standard Oil of Indiana. The chapter ends with a discussion of the new activists in safety and health.

Business-funded institutions were the only organized constituency in occupational safety and health until the end of the 1960s. Organizations such as the National Safety Council (NSC), the National Council on Compensation Insurance (NCCI), the American National Standards Institute (ANSI), the Industrial Health Foundation (IHF), the American Occupational Medical Association (AOMA),[1] the American Industrial Hygiene Association (AIHA), and the American Conference of Governmental Industrial Hygienists (ACGIH) constitute the heart of the traditional compensation-safety apparatus. Differences in philosophy between these organizations have been few, even where their membership is made up of governmental rather than private employees; and they rarely express public disagreement with the positions on health and safety taken by the National Association of Manufacturers or the U.S. Chamber of Commerce. In the fierce conflicts over standards-setting and enforcement under OSHA the traditional groups have often been bypassed by consulting firms or by trade groups such as the Manufacturing Chemists Association or the Asbestos Information Association on issues which directly affect their respective constituencies.[2] The doctrines and modes of action developed by the traditional organizations, however, still dominate business activity in health and safety. These organizations share common interests in that they are funded by big business and insurance sources; they believe in voluntary action by the private sector rather than government regulation of working conditions; they support common ventures and do not criticize each other publicly; they hold similar positions on those issues on which they take a stand; and they claim to be disinterested and scientifically neutral between management and labor. All of them subscribe to a common set of doctrines (amounting to an operational code) about the causes and solutions for health and safety problems,[3] and share industry's concern with keeping its costs to a minimum. In congressional testimony prior to the passage of the OSHA law these organizations uniformly supported weak health and safety bills.

Indoctrination and the National Safety Council: How Safe Is "Safety First"?

The importance of shared doctrine cannot be overemphasized, especially in a policy area like occupational health and safety, where business defined the problems without challenge for decades. In their effect on the labor force, management safety doctrines constitute a form of ideological counterinsurgency which shifts the blame for accidents and diseases away from management. Instead of problems of industrial design and speedup, accidents become inevitable problems of a supposedly immutable human nature. The effect is to induce passivity about working conditions. It would make sense to blame workers for accidents if they controlled factory design and the organization of work, but those questions are considered to be management prerogatives. In the present context, safety doctrines which blame employees tend to become everyday common sense until they are exposed and challenged from the worker's perspective.[4]

The following propositions are a fair statement of management compensation-safety doctrines at present; none of the organizations we have mentioned has disagreed with more than one of the seven doctrines in testimony on the various job health and safety bills before Congress: (1) most accidents are caused by worker carelessness rather than by faulty plant and equipment design and production speedup;[5] (2) the voluntary education of management and workers to better safety practices rather than the strict enforcement of government regulations is the time-proven and effective method of preventing accidents;[6] (3) health hazards at work have been controlled by most companies and are of declining importance, and new substances are carefully checked for danger before worker exposure to them;[7] (4) workplace standards should be developed by "standard-making bodies in the private sector";[8] (5) government regulation should be on the state rather than the federal level, and enforcement should be cooperative rather than punitive;[9] (6) safety pays, thus there is no need for safety and health regulations because the high cost to employers of injury, disease, and death forces them to take all feasible precautions anyway;[10] (7) the industrial safety movement as represented by organizations such as the National Safety Council has been a success for decades and there is no reason for basic changes in how it operates.[11]

The oldest and most important organization in the compensation-safety apparatus is the National Safety Council, which was founded in 1912 as a response to the passage of workers' compensation laws and the unrest over working conditions. Large corporations often found that by enforcing rudimentary safety practices and installing simple safety equipment, they could save themselves money in accident costs. For example, over an eight-year period at the beginning of this century, U.S. Steel spent $5 million on safety devices and education. As a result the accident rate fell 40 percent, with a 35 percent saving in accident-related costs.[12] After some early successes in reducing accident rates, the corporate safety movement seemed to run out of steam. Compensation costs were stabilizing, and the Great Steel Strike and the red scare ended most union militancy and radical activity. The return to normalcy began a forty-year consignment to oblivion of the safety issue.[13]

Today the National Safety Council operates on an annual budget of about $13 million, sometimes bolstered by federal grants. Half of its program is devoted to occupational safety and health, and twenty-four of its thirty-one trustees are managers at large corporations. The Reverend Billy Graham is the only trustee widely known to the public.[14] The organization heavily advertises and promotes safety in all spheres of life around the country, and is best known to the general public as the stern prophet of holiday weekend death tolls on the highway. In its safety propaganda the NSC has focused blame on the careless worker and on the "nut behind the wheel" rather than on corporate responsibility for dangerous designs and practices.[15] It has sponsored little or no serious accident research, and its statistical work has mainly served to understate the number of injuries and deaths, while totally ignoring job-related illnesses. The National Safety Council's attractively illustrated reference works, while dealing competently with job safety, largely discount the dangers of occupational disease. Its *Catalog-Poster Directory 1971–1972* lists over 100 films and over 1300 posters on safety subjects. Most of the films are made by trade associations and corporations. According to Robert Fowler, former safety committee head, International Association of Machinists (Local 1781), who screened many hours of them, the films are boring and condescending to workers.

Workers' Compensation and the National Council on Compensation Insurance (NCCI)

Within the apparatus there has been a partial division of effort between insurance interests, primarily concerned with workers' compensation, and corporate-sponsored organizations, such as the NSC, that are more concerned with workplace standards, safety publicity, OSHA, and accident prevention. The American Insurance Association (AIA) and the American Mutual Insurance Alliance are the principal lobbying representatives of the insurance business on both state and national levels in relation to workers' compensation. In recent years they have devoted their energies to promoting the cosmetic reforms proposed by the insurance-dominated National Commission on State Workmen's Compensation Laws in order to prevent federal regulation. In 1973 they helped defeat attempts to start state workers' compensation funds in the states of Hawaii, Maine, and Montana.[16] Although figures on the American Insurance

Association's annual budget were not made available to me, it must be over $10 million, judging from the dues paid by some companies. For example, the Aetna Insurance Company of Hartford and the Fireman's Fund Insurance Group, which together took in 6.6 percent of all property-liability premiums in 1975, paid a total of $585,521 in dues that year. The informational base for lobbying has been provided since the early 1920s by the insurance business-supported National Council on Compensation Insurance and local groups such as the California Inspection Rating Bureau, a $4.3 million dollar a year operation which just celebrated its sixtieth anniversary.[17] It is clear that groups such as the NCCI are very important in setting compensation policy, but, as with much of the insurance business, there has been little independent scrutiny of their activities.

Safety Standards and the
American National Standards Institute (ANSI)

Most OSHA safety standards were adopted directly from ANSI. In 1969, before the passage of OSHA, the Department of Labor requested that ANSI develop 83 "critically needed" standards, and in May 1969 adopted 180 ANSI standards as official Department of Labor standards. These were taken over intact under Section 6(a) of the OSHA law. Where ANSI's standards conflicted with more stringent regulations developed by other groups, as in the area of airborne contaminants, ANSI's received precedence in OSHA regulations.

ANSI is an umbrella organization that serves to legitimize standards developed by private industry in many areas besides occupational safety and health. Indeed, it calls its standards American National Standards, as if they had been promulgated by an official government body. In the late 1960s, 100 large corporations, 160 trade associations, 6 trade unions, and a number of government agencies were authorized to participate in the deliberations of ANSI's committees. At that time, support for the group's activities came from over $900,000 in dues paid by large corporations and from the $25 million operating budgets of the 3,000 technical standards committees, whose 36,000 members work on company time. But even these figures understate the money spent on private standards development; as ANSI's managing director, Donald Peyton, pointed out in the 1970 Senate hearings: "Budget figures include only the amount spent on internal administrative overhead and staff travel. Separate would be the tremendous amount of money spent by members of technical committees on travel, living expenses and their individual salary expenses plus the much greater cost of research programs undertaken in the private sector as a basis for standards."[18] Peyton added that about one-quarter of this massive effort is devoted to occupational safety and health standards. By contrast, in 1972 OSHA and NIOSH together spent a total of only $3.1 million on occupational safety and health standards development, though the amount was expected to triple.[19] Basically, ANSI incorporated the government's efforts. In the words of ANSI's president:

"ANSI is not an organization in competition with the government. On the contrary, the competence of the personnel within the government agencies is an integral part of the effort organized by ANSI, its member bodies, and its councils."[20]

ANSI standards tend to be lenient toward business. The work-injury measurement standard, for example, was originally created by government statisticians and later it was taken over by private interests. Until the passage of OSHA it was the only measure of work-injury rates in the United States. After its private takeover in the 1920s, it was continually redefined so as to understate the incidence of work injuries, and had no category for occupational disease at all. The committee that devised ANSI's "USA Standard Method of Recording and Measuring Work Injury Experience—Z16.1" includes among its forty-seven organizations thirty-three business and insurance trade associations and fourteen governmental organizations. The AFL-CIO was allotted only one place on the committee, and did not even bother to appoint a representative.[21]

ANSI's managing director admitted that labor's membership on standards-setting committees is solicited in order to add a gloss of legitimacy to the proceedings. In regard to labor's role in setting a longshoring standard, he blamed labor for not participating:

> On the matter of union representation we make every effort to obtain the participation of affected parties. Since this is a voluntary system, we have no control over whether or not they respond affirmatively or fully participate in the work. Nonetheless, you will note from the enclosed memorandum that the U.S. Department of Labor Bulletin lists only four unions involved in stevedoring and warehousing. The fact that they were invited should be ample evidence that representation was sought. The fact that only two were still active at the conclusion of the project is a matter of choice of the unions, not the Standards Institute.[22]

Health Hazards and Health Standards

During World War II the U.S. Public Health Service carried out its first attempts to determine dangerous levels of exposure to airborne contaminants. By 1946 a list of the "maximum allowable concentrations" (MACs) for a number of airborne contaminants was made generally available, though the MACs were not legally enforceable. They soon proved to be inadequate as a measure, because they could not "reflect the relationship between concentration of a potentially harmful substance and duration of exposure." As a substitute for MACs the ACGIH invented the "threshold limit value" (TLV). This eight-hour time-weighted average is defined as "the concentration of an airborne contaminant to which workers may be exposed repeatedly day after day, without adverse effects." It presumes that for all substances such a concentration can be determined, usually by extrapolation from toxicity tests in animals and observation of exposed workers. In setting the TLVs a substance is considered acceptable if exposure at the TLV level causes physiological changes in the exposed worker but the changes are not harmful.[23]

As a result, the TLV figures generally have been high enough to harm workers. Permitted industrial exposures to carbon monoxide, nitrogren dioxide, and sulfur dioxide are four to forty times as high as exposures recommended by the Environmental Protection Agency for air in the environment at large which workers have to breathe after returning from a day at a dirty job. The TLVs presume that the people at risk in the industrial environment will be healthy young men, rather than women of childbearing age and older people who have already suffered substantial damage to their health. The TLVs were almost all set with immediate toxic effects in mind, rather than considering the probabilities of causing cancer, birth defects, or genetic damage, or the problems from inhaling mixtures of industrial poisons.

Since World War II there has been an explosion of new chemicals into the workplace and the environment. The number of commercially available chemical compounds rose from 17,000 to 47,000 between 1958 and 1971.[24] Since there are only 500 TLVs in the OSHA regulations for existing hazardous substances, thousands of known and suspected toxic chemicals lack exposure standards. The adverse health effects of radar and other microwave uses, developed by the military, were ignored or kept secret for decades. Since many of the new hazards cause health damage which manifests itself only decades or generations later, the worst is probably still to come. Already scientists in the USSR and the West are concerned about a steep rise in the incidence of birth defects, and many believe that the increase in the cancer death rate owes a great deal to the proliferation of new chemicals in the workplace and the environment at large.[25]

The Kepone case illustrates the problem posed by the huge number of new substances. In Hopewell, Virginia, the Allied Chemical Corporation set up a puppet business, the Life Sciences Products Company, to produce Kepone, an insecticide for banana growers in Latin America, Asia, and Africa; from their own research Allied knew the product was so deadly that the company was afraid to produce it under its own auspices. The workers were told nothing of Kepone's dangers when they began to produce it in an old garage near the Allied Chemical plant, and there was no OSHA standard regulating exposure to the substance. Although Allied initially denied responsibility, a federal court forced the company to pay record damages of $13.5 million, after waste Kepone had poisoned an entire river valley and wiped out a flourishing fishing industry on the James River. Now pending is a $100 million damage suit by former employees, many of whom have suffered permanent bodily "shakes," nerves, and sterility. The effect on future generations is still uncertain, and, unfortunately, Kepone is far from being the only case of its kind.[26]

Asbestos, Corporate-Sponsored Research, and the Setting of Health Standards

With the passage of the new federal health and safety law, responsibility for setting health and safety standards shifted from the private business groups such as ANSI to OSHA in the Department of Labor. Standards-setting now requires NIOSH to review the relevant scientific literature and recommend a standard to OSHA, which then hold public hearings at which interested parties can state their cases. Health has usually overshadowed safety. In its first four years, under a Republican administration, OSHA was reluctant to recommend new health standards, so most of the action occurred after union and public interest lawsuits. Large battles have already occurred, pitting industry against labor and its allies over new standards for asbestos, carcinogens, vinyl chloride, power press safety, coke ovens, noise, and lead.

Standards hearings, like any other regulatory procedure, consume a tremendous amount of time, money, and expertise. Testimony is generally in the mysterious languages of medicine, economics, and the law, and extremely complete documentation usually is required. Preparation of a case can cost hundreds of thousands of dollars. Since hearings usually occur in Washington, D.C., individual workers and local unions are often excluded from the proceedings. Despite the expenses of standards-setting under OSHA, the relative openness of the process has allowed workers and their representatives a much greater voice than in the past.[27]

The long and continuing battle to control asbestos illustrates the difficulties in securing and enforcing protective regulations in occupational health. Even when the scientific evidence is unequivocal, it is necessary to fight against the fear that workers have of losing their jobs in an economy with a permanently high level of unemployment. The asbestos story also shows the difficulty in securing honest research results about a profitable but hazardous commodity such as asbestos, which, according to Dr. Irving J. Selikoff, will kill two-fifths of those exposed to it throughout their working lives.

Some awareness of the hazards of asbestos has existed for thousands of years. As early as the first century, Roman naturalist Pliny the Elder mentioned a lung disease suffered by slaves who mined asbestos, and described the makeshift respirators they devised to protect themselves.[28] In 1906 the first modern medical account of a worker's death caused by asbestos was reported by a British physician.[29] By 1918 U.S. and Canadian insurance companies had stopped selling personal life insurance policies to asbestos workers, although there were still only a few thousand such workers scattered around the continent. After World War I the asbestos industry in North America began a long-term expansion, which has not ceased to this day. Between 1925 and 1974 the leading corporation in the field, Johns-Manville, increased its annual sales from $40 million to over $1 billion, as thousands of new uses were developed for asbestos. Today in the United States 90,000 people work directly with asbestos and another 5 million work with asbestos-containing products every day. Workers bring asbestos home on their clothes, thereby exposing their families, and, in fact, asbestos use is so common that most Americans have asbestos fibres in their lungs. The profits and jobs linked to asbestos, however, have made it politically difficult to cut back its use.

Since the 1920s Johns-Manville and the industry as a whole have known about the hazards of asbestos use. The industry decided to deal with the problem by funding medical research that would discredit reports of asbestos hazards and "keep . . . a check on workers' health while telling them as little as possible," according to Dr. David Kotelchuk of the Health Policy Advisory Center. In 1929 the asbestos industry hired the Metropolitan Life Insurance Company to conduct a study of employees from five plants and mines in the United States and Canada, mostly Johns-Manville facilities. Completed in January 1931, the results were not published until 1935, as a Public Health Service report. Although the study showed that 67 of the 126 people examined had asbestosis, the author of the study, according to Kotelchuk, "simply listed the number of workers in each category and hurried on without comment. Short of suppressing the data, they could have done no less." The same year the Metropolitan Life study was published, cases of lung cancer associated with asbestosis were reported in Britain and the United States. The Report of the Chief Inspector of Factories of Great Britain for 1947 showed a higher than expected incidence of lung cancer in individuals who had died of asbestosis, and in 1955 it was proven unequivocally that exposure to asbestos causes lung cancer. Two research laboratory scientists partially funded by Johns-Manville promptly disputed the asbestosis-lung cancer connection, without bothering to cite contrary evidence.[30]

By 1960, sixty-three scientific papers on the problem of asbestos exposure and health had been published in the United States, Great Britain, and Canada. The fifty-two papers published independently of the asbestos industry showed asbestos to be a dangerous source of asbestosis and lung cancer; the eleven sponsored by industry presented virtually the opposite conclusions, rejecting the connection between asbestos exposure and lung cancer and minimizing the seriousness of asbestosis. The independent medical researchers who wrote about the hazards, while reflecting a "humane concern for the afflicted workers," generally buried their findings in technical publications, inaccessible to the general reader. Neither workers nor the public (which is also subject to asbestos-induced cancer) took an active part in decisions about how asbestos was to be handled. Those decisions were left with the asbestos industry and a compliant government.[31]

Only in the early 1960s did Dr. Irving J. Selikoff and his associates at the Mt. Sinai School of Medicine in New York City succeed in tracing the effects of asbestos exposure in workers without having to rely on corporate records. Selikoff, who already had earned a reputation as a brilliant lung specialist, employed his political and public relations talents to bring the truth about asbestos to the public at large. Of equal importance were his group's extensive educational sessions among asbestos workers, where he and other scientists learned about working with asbestos and how to explain scientific results in language lay people could understand. By the late 1960s members of the Asbestos Workers Union all over the country had an understanding of the dangers to their health. Research money flowed to Mt. Sinai from the Health Research Council of the City of New York, the U.S. Public Health Service, and the American Cancer Society. In addition, the Asbestos Workers gave a sum (matched by Johns-Manville) to obtain equipment for investigation on safer methods and work practices.[32]

Mt. Sinai's efforts with the Asbestos Workers blew the lid off the asbestos situation, and played an important part in creating the political climate which led to the passage of the Occupational Safety and Health Act of 1970. By then most asbestos researchers not in the pay of industry were agreed that asbestos was a potent carcinogen and that there was no level of exposure that did not involve at least some risk of contracting cancer. These scientists usually concurred that any exposure to over two fibres—longer than five microns, averaged over an eight-hour day—per cubic centimeter of air over a working lifetime would lead to some incidence of asbestosis, a progressive disease which scars the lungs and prevents adequate transfer of oxygen to the blood; some even went so far as to advocate the complete elimination of asbestos use.[33]

Since denials or evasions of the harmfulness of asbestos were no longer credible, the industry was forced to reevaluate its research strategy, greatly increasing spending on scientific research. Since 1960 it has sponsored the publication of three times the number of scientific papers on asbestos as it did in the thirty preceding years. In 1972 the asbestos industry spent $8.5 million on "research and development," much of it on health matters, compared to the $260,000 spent by NIOSH on asbestos research. By spreading its grants widely in a period of declining support for most medical research, industry's idea was to keep direct criticism at a minimum. As the executive director of the Asbestos Information Association noted, "it would be extremely difficult to find a credible researcher in the country whose work in asbestos has not been or is not presently supported, at least in part, by the asbestos industry."[34]

Industry-funded studies still try to minimize the health problems related to asbestos. For example, Johns-Manville physicians and executives have tried to blame asbestos-induced cancer on amosite asbestos, rather than on the chrysotile type which the company produces almost exclusively. Another industry ploy is to design studies that examine a large sample of employees working at many different jobs, which "tends to bury the effect of unhealthy subgroups." The summary and conclusions point to the similarity of the large group's sickness and death rates to those of the general population, dismissing the fact that workers usually are healthier than the population as a whole.[35]

As the Tyler, Texas, asbestos scandal at an insulation plant owned by the Pittsburgh-Corning Corporation showed, the new Nixon-appointed OSHA administration did not take seriously the violations of its own health standard.[36] For years, government surveys had indicated extremely high asbestos dust levels at the plant. Dr. William M. Johnson of NIOSH, for example, described his first impressions as follows:

> The place was an unholy mess. Why, compared with it, the Port Allegheny plant looked like a hospital operating room! A thick

layer of dust coated everything—from floors, ceilings and rafters to drinking fountains. As we walked through the interior, we saw men forking asbestos fibre into a feeding machine as if it were hay. They obviously had no idea of the hazard involved. Farther down the line, we came upon some fellows with respirators hanging around their necks, who were sitting in an open doorway eating watermelon. I remember turning to Dr. Richard Spiegel, one of my assistants. "This is intolerable," I told him. He was as shocked as I was.

Since the Tyler plant had been operating for only seventeen years Dr. Johnson didn't expect to find cancer, but seven of the eighteen workers with more than ten years of employment at the factory showed unmistakable signs of asbestosis.

Forty-two of the fifty-four air samples taken during an OSHA inspection exceeded the existing standard of twelve fibres per cubic centimeter. Pittsburgh-Corning was cited for such violations as improper wearing of respirators, failure to examine workers to determine whether they had the physical capacity to wear respirators, inadequate housekeeping, and insufficient dust control. No mention was made in the OSHA citation that the Tyler plant had exceeded the legal limit for exposure to asbestos dust. The total fine for these "nonserious (other) violations" came to $210.

Partly as a result of the Tyler situation, organized labor and its allies began to pressure Secretary of Labor Hodgson to declare a two-fibre standard for exposure to airborne asbestos in the first test of OSHA's public standards-setting procedures. Hodgson, caught between the asbestos industry and labor, declared a temporary standard of five fibres (down from twelve fibres), appointed a five-person Advisory Committee on the Asbestos Standard, and called for hearings to begin on a permanent standard in March 1972. The hearings demonstrated the ability of the independent scientific community, when backed by labor, to resist manipulation by industry: both the Advisory Committee (representing industry, labor, government, and science) and the NIOSH criteria document called for a two-fibre rather than a five-fibre standard. Only industry-funded researchers recommended a five-fibre standard, and the twelve-fibre standard was no longer taken seriously by anyone. Asbestos industry executives claimed they would have to shut down U.S. plants if a two-fibre standard were passed.

In the face of resistance by university and NIOSH researchers to industry regimentation, OSHA hastily funded a survey by the Arthur D. Little consulting firm which asked its (mostly pro-industry) respondents about the probable health effects of exposure over a working lifetime to two, five, twelve, and thirty asbestos fibres per cubic centimeter of air, and the immediate cost to industry of reducing exposure to given levels. The costs and lives of workers and the public which would be saved were not even considered, and such economic calculations were to become a permanent part of standards-making. Clearly, as Sheldon Samuels of the AFL-CIO's Industrial Union Department pointed out, "the executive branch of government has decided on its own that the cost to the employer of meeting any new occupational health standard must fall within an economic range that is acceptable to industry."[37] Ultimately, the exposure standard was set at five fibres per cubic centimeter, to be reduced to two fibres by July 1, 1976. That standard successfully withstood an industry lawsuit, and in late 1976 NIOSH proposed a further reduction to 0.1 fibres, in order to sharply reduce the risk of cancer.[38]

Industry response to the pressure over asbestos has varied. Pittsburgh-Corning sold its asbestos insulation plant in Tyler, Texas, shipping the usable machinery back to headquarters and selling or burning the rest of the plant, and Pittsburgh-Corning and Johns-Manville and some industry trade associations were sued for $100 million over their failure to warn and protect the Tyler workers. Many firms, such as Raybestos-Manhattan Corporation, are considering moving to Mexico, Taiwan, or South Korea, where worker exposure to asbestos is without legal restriction.[39]

Of the 689 workers at Johns-Manville's Manville, New Jersey, plant who were studied from January 1, 1959 through December 31, 1971, 96 died of asbestos-related diseases, compared to an expected total of 28.[40] By the spring of 1973 over four-fifths of the plant's dust monitoring stations recorded levels of two fibres or less. Robert Klinger, vice-president of United Papermakers and Paperworkers Local 800 at the Manville plant, explained why the levels came down:

> We had a long and costly strike in the autumn of 1970. As part of the settlement, the company guaranteed to make a real effort to reduce dust levels in the plant. Then, too, in 1969 Johns-Manville paid out nearly nine hundred thousand dollars in workmen's compensation in New Jersey for asbestosis alone, over and above what it may have settled out of court in litigation brought against it by workers, or families of workers, who had contracted asbestos-induced cancer. In addition, the work of men like Dr. Selikoff and Dr. Hammond [of the American Cancer Society] was by then piling proof upon proof of the association between asbestos and disease.
>
> Since April of 1971, when a government survey showed that some dust counts in the Manville textile operations were running as high as twenty fibres per cubic centimeter, there has been considerable improvement ... the J-M people simply saw the writing on the wall, and decided they had better act.[41]

At other plants around the country Johns-Manville faced protests, strikes, and lawsuits over asbestos policies, with the result that some of the plants have been cleaned up a great deal.[42] Even so, the company's president has criticized the "unnecessarily harsh" OSHA rules, complaining that "we have suffered unduly at the hands of people who make their living with health scare stories. ... This is an industrial hygiene problem, not a problem of the public."[43]

Because of the risk to workers and the public at large, many states and municipalities have forbidden the open spraying of asbestos insulation, and construction workers have told me that asbestos is now being replaced by fiberglass as an insulator. Smaller companies, however, which are at a comparative disadvantage in instituting work environment controls, have largely ignored the regulation if asbestos is not easily substitutable. Many have not even heard of it. Others have taken preventive measures only because of union initiative.[44]

Third-party liability suits by asbestos insulators and pesticide formulators against manufacturers have become rather common. Asbestos lawsuits are generally for a million dollars or more for each asbestos-caused death, and Johns-Manville admits to 271 suits on asbestos disease, with many multiple plaintiffs. The Tyler, Texas, suit was settled for over $14 million: $8.1 million from Pittsburgh-Corning (jointly owned by PPG Industries and Corning Glass); $5.7 million from the federal government; and $1 million from the former owners of the plant. The union itself, the Oil, Chemical, and Atomic Workers, agreed to pay $150,000 into the settlement to avoid a potentially more costly lawsuit by PPG Industries, which claimed that the union was negligent in not informing its workers sooner. The money went to 445 workers and their descendants who brought the suit. Already, around twenty-five deaths and seventy-five to eighty cases of asbestos-related disease have been attributed to working conditions at the plant. Naturally the asbestos and pesticide companies are extremely nervous over the huge costs of the lawsuits, and rumor has it that unions which are sued by their members are eager to settle out of court, to avoid the expense and public embarrassment of their neglect to take action.[45]

The Misuse of Medical Research and the Denial of Health Hazards: Beryllium, Diatomaceous Earth, PCBs, and Cotton Dust

The denial of the health-destroying effects of poisons used in industry and the concealment of information about their toxic effects from the medical-scientific community and workers have been used for decades by industry-oriented scientists. Exposure to beryllium, a metal used in rocket fuel and as an alloy, provides an example. Dr. Harriet L. Hardy, trained by Dr. Alice Hamilton, was a pioneer researcher of the noxious effects of beryllium metal, and founded the Massachusetts General Beryllium Case Registry to enable her to study, treat, and prevent beryllium disease. She wrote of her findings:

> Unfortunately, by no means all cases of beryllium disease are known to the Registry, not even all those so diagnosed. . . . The help of over 300 physicians who have sent case records to the Registry is gratefully acknowledged as making what I have to offer of real value. In contrast, with rare exceptions, industry and insurance companies withhold data on occupational disease—its character and incidence. This fact has great influence on the acquiring of knowledge of industrial illness in other as well as the beryllium-using industry in the United States.[46]

For almost twenty years industry and the Atomic Energy Commission had claimed worker exposure to beryllium was harmless, based on an erroneous bulletin which the Public Health Service published a year after the death of the first patient identified by Dr. Hardy's registry as suffering from chronic beryllium poisoning. Dr. Hardy charged that acceptance of this claim was enhanced by the activities of medical professionals: "A few consultant doctors and industrial hygienists, by their publications, talks at professional societies and appearances in court, appear to have been used by some members of the beryllium industry to further what are considered legitimate economic ends." Her conclusions placed responsibility for beryllium poisoning with the private industrialists:

> Twenty years of study of beryllium poisoning have forced me to attempt to assess the economic influences that affect the accumulation of accurate knowledge and all too often the care of a sick worker. There is nothing in the physician's training to prepare him to judge such influences. . . . Many physicians and research workers . . . fail to realize the possible hardships to patients or investigation that may arise when the industrialist puts profit first and public responsibility last.

Another example of the concealment of information related to occupational health concerns the harmful effects of diatomaceous earth often used as a filter or as insulation. The harmful effects were first described in the U.S. medical literature in 1932. As early as 1938 physicians in the industry had collected clinical and pathological materials which provided "convincing evidence of the existence of a disabling occupational lung disease. Even so, in 1940 the chief of the industrial hygiene division of the state of California was barred from inspecting the diatomaceous earth mines and mills owned by Johns-Manville. It was not until 1946 that company physicians would admit that diatomaceous earth could produce a disabling and sometimes fatal disease. A few workers began to receive compensation after difficult legal battles, but conditions improved substantially only after a seven-month strike on the issue of dust exposure by a local of the International Chemical Workers Union in 1953. And yet employees at the same operation only recently found out that Johns-Manville has been killing them for years by exposure to asbestos dust.[47]

One ploy is to simply call poisoning by a different name. At the National Conference on Cotton Dust and Health, for example, Dr. Robert T.P. de Treville of the Industrial Health Foundation refused to call byssinosis (a lung disease caused by exposure to cotton dust) a "disease," in order to keep workers from worrying about it. He felt it was

> best described as a "symptom complex" rather than a disease in the usual sense. We feel that this term may be preferable, first, in order not to unduly alarm workers as we attempt to protect their health; and secondly, to help avoid unfair designation of cotton as an unduly hazardous material for use in the textile industry, raising the fear the engineering control of it may be costly and that it may be better, therefore, to switch to some less costly material.[48]

The intention to protect cotton manufacturers' profits at the expense of the workers' health could not be clearer.

The problem is not only one of deliberate concealment,

however. Ecologist Barry Commoner has pointed out how little interest and prestige the medical-scientific world gives to the problems of workers in his discussion of the history of exposures to polychlorinated biphenyles (PCBs), substances chemically similar to the insecticide DDT which are used as industrial coolants: "Although the hazard from PCBs was first discovered in the work place, in 1933, the problem was given relatively little scientific attention (beyond clinical description of chloracne [a skin disease] and its association with PCBs) until 30 years later, when it was first recognized as an *environmental* hazard." Commoner quotes a finding by J. G. Vos that when they were first used, "little work was done on the toxicology of the PCBs, and this was only in relation to the risks of occupational exposure.... Many more studies were made as soon as it appeared that the extremely stable PCBs became a threat to the environment and its wildlife, and accidents occurred of acute poisoning in man and animals."[49] Monsanto, the sole U.S. producer of PCBs, took action to cut environmental exposure only after an uproar in the early 1970s.

The Captive Specialties: Occupational Medicine, Industrial Hygiene, Industrial Nursing, and Safety Engineering

The instruments of corporate domination of the occupational health and safety specialties have included the control of labor recruitment, job training, and the prestige of these specialties. The lack of a labor-oriented "counter-establishment" until recently and the shortage of funds outside of corporate control to investigate job-related health problems has made it difficult for worker advocates to make a living in the health and safety field.

Within the formal safety and health specialties it is not surprising that physicians, with their greater prestige and ability to make money in a variety of settings, were the chief early dissenters from corporate modes of thought. From the virtual founding of occupational medicine in the United States by the great Alice Hamilton[50] to the present, there have always been a few physicians willing to speak out for the worker. By contrast, industrial hygienists and safety engineers, like engineers in general, have become more dependent on big business and subservient to it. Not a single industrial hygienist, nurse, or safety engineer disagreed publicly with the pro-industry positions of their professional societies in testimony preceding the passage of the OSHA law of 1970; the public dissenters were all physicians in research or private practice who were immune or indifferent to direct economic pressure from companies. Since OSHA there have been more dissenters from the other specialities. Yet except for a few self-taught workers, union leaders and employees, and impoverished activists, working as a health and safety specialist still means being an employee or consultant for a large corporation, insurance company, or earning a small salary in a forgotten government office.[51]

Occupational Medicine

The medical field as a whole has paid very little attention to the problems of occupational health and safety. To begin with, little is taught about it in medical school. The class of 1968 at Harvard Medical School, for example, received only one lecture on the subject in four years of training, compared to twelve lectures in 1949[52] and "to this day medicine is taught in the schools as though the industrial revolution had not yet occurred."[53] Few medical students or young doctors ever actually see the ravages of acute work injury or disease, however, because since the advent of the workers' compensation system most of those hurt on the job have become the captive clientele of full-time company doctors, physicians on insurance company panels, or private practitioners located in industrial areas. Finally, the ghettoization of industrial medicine has been enhanced by the influence of big business on medical school boards of trustees, which naturally neglect research into dangerous conditions in industry.[54] As a result, few physicians ask their patients about their conditions of work: a poll taken by Art Button of Teamsters Local 688 in St. Louis showed that physicians working for the union-controlled Labor Health Institute were no more likely than other doctors to try to relate illness to dangerous working environments.[55]

Doctors' attitudes toward occupational illness have served to reinforce its neglect. Most doctors are recruited to industry from private practice and start out with the anti-worker attitudes common to their class background. According to Dr. Irving R. Tabershaw, editor of the *Journal of Occupational Medicine*:

> The physicians' biases . . . are no different from those of their peers in their socio-economic class. Since most physicians come from the middle class or soon attain that status, they accept that somebody must do the dirty and dangerous work, but it is to be avoided by their immediate family and friends, if at all possible. Physical risk may be sought voluntarily in sports but not as a means of livelihood.[56]

In addition, the low prestige among medical students and doctors of industrial medicine is a consequence of its primary social role: to keep down workers' compensation costs. In most states, workers injured on the job must go to a doctor chosen by the company, or else pay the medical bills. Company doctors have the legal responsibility of determining whether the injured are fit to return to work; people who stay away from work without the doctor's permission are subject to firing. Moreover, in compensation proceedings a physician's testimony determines whether or not disability benefits are granted, especially in costly and often contested permanent disability cases. Naturally, the company doctor has to testify on behalf of the company. In large corporations the company physician also has the task of screening out unhealthy job applicants, particularly those with back problems, who might be subject to costly injuries.[57]

For most physicians, becoming a company doctor is more a matter of coming to share the profit-making orientation of management than of acquiring new technical skills such as the ability to assess the hazards of the work environment. Dr. Clarence D. Selby's "Studies of the Medical and Surgical Care of Industrial Workers," written for the Public Health Service, described the company doctor's place as it was in 1919:

> Although fundamentally the science of medicine, the position which industrial medicine occupies is similar to that of employment, safety, and compensation. All are specialties in the science of management.
>
> Physicians who do not understand this relationship (and medical training does not necessarily contribute to this understanding) have reluctance in accepting the materialistic viewpoint of employers, and, conversely, have difficulty in persuading employers to accept their professional points of view.[58]

Not surprisingly, Selby found that in only 4 percent of the plants surveyed was the physician ". . . a real medical director, with freedom, responsibility, and authority to which he was entitled." In his Bulletin No. 99, for example, Selby pointed out that

> examinations for employment, as usually made, were exceedingly superficial, amounting to little more than inspections for obvious defects, and although the avowed purpose was properly to place employees, the real purpose in some places seemed to be the exclusion or the recording of defects that might subsequently complicate injuries or become involved in claims for compensation.[59]

The fact that a number of firms did not require examinations in cases of scarce labor, as Selby pointed out, would suggest that the primary purpose of the examination was the "exclusion of defectives."

The use of the preemployment physical today has changed little, according to Dr. Tabershaw:

> Unfortunately, in the past (and perhaps still in the present) the preemployment examination was used to weed out undesirable Workmen's Compensation risks or as a subtle technique for dis-

crimination in employment. Until recently, the medical examination was not given . . . to alert the worker to the risks of the employment he was seeking. It was management that was evaluating the risk of hiring a particular worker, even though in some instances the person also benefitted. . . .[60]

Since the structure of workers' compensation has hardly changed in fifty years, the role of the occupational physician has not changed either. In 1961 Dr. William P. Shepard, former chief medical director at Metropolitan Life and former head of the Council on Occupational Health, American Medical Association, wrote *The Physician in Industry*, in order to help the industrial physician "to orient himself in a new environment." On page one he made it perfectly clear that "the physician's place in the industrial system is quite different from that to which he has become accustomed in private practice." This means "he is not top man as he is in the hospital or his private office. His services are strictly ancillary to the main purpose of the business: production at a profit. His value depends upon his willingness and ability to work with others to achieve that main purpose."[61]

As a response to public criticism of occupational medicine, company physicians have begun to defend themselves in professional journals.[62] In addition, the American Occupational Medical Association has approved a "code of ethical conduct" for its members. While its expressed principles are praiseworthy, the code still denies employees direct access to their medical records, ignores workers' compensation, and fails to accord workers the right to choose their own physicians or determine who should run industrial medical departments—as though these questions were irrelevant to the faults of occupational medical practice. Struggles over just those issues have ocurred at United Auto Workers Local 6 near Chicago, an International Harvester plant.[63]

Industrial Hygiene

Industrial hygienists are trained to prevent occupational disease by detection, measurement, and removal of threats to health from the work environment. At least four universities in the United States have doctoral programs in which industrial hygiene is a major or important part of the course of study.[64] Industrial hygienists are grouped into two major societies: the American Industrial Hygiene Association (AIHA), with 1,600 privately employed members,[65] and the American Conference of Governmental Industrial Hygienists (ACGIH), with about 1,100 members in the late 1960s. Forty-one percent of ACGIH's members work for state and local government; 25 percent for the military; 21 percent for the federal government; and 1 percent are connected primarily with universities.[66] The AIHA, which is backed by the private sector, is the stronger organization of the two. It publishes its own journal and rather expensive brochures, which make the ACGIH's publications look spartan by contrast.

Both societies supported weak industry-backed bills in congressional testimony about the safety and health acts, but there are important historical differences between the two.[67] The ACGIH split off from the AIHA in 1938 expressly to keep out corporate influence. Until 1943 it held its annual convention separately from the AIHA. A prominent hygienist, Dr. William G. Frederick (winner of the annual ACGIH "meritorious achievement award" in 1968) commented on industrial hygienists employed by private industry:

[They] tend to have their primary loyalty to their employer and have rather consistently . . . and aggressively opposed steps forward to try and improve the health of the worker. That has been a very real thing and I can assure you if the ACGIH had not been primarily worker oriented rather than employer oriented, the Threshold Limit Committee would never have come into existence and been as productive as it has.[68]

Dr. Frederick's comments came in the context of his opposition to opening membership in ACGIH to industrial hygienists working full time for private companies which were completely funded by the government, mostly in military production.[69]

However, the line between the ACGIH and the AIHA is more apparent than real. The two organizations jointly certify candidates through the American Board of Industrial Hygiene; they hold joint national meetings annually, and they participate on many committees together. Members of ACGIH working for industry-financed organizations often chair important committees which have delayed the recognition of the dangerous qualities of materials like coal dust. Other factors that work for the interpenetration of the governmental and private sectors in industrial hygiene are the higher salaries in private employment[70] and the constant movement of people between the two sectors. This heightens an identity of values which reflects the interests of the stronger sector.

Attempts by activist industrial hygienists since 1975 to enact a code of ethics which strongly emphasizes worker protection were defeated by the American Industrial Hygiene Association (AIHA). Rather than recognizing an overall responsibility of the industrial hygienist to "protect, promote, and defend the health and safety of the individual at the workplace," the code of ethics approved by AIHA's law committee merely exhorts the industrial hygienist to practice in an ethical manner. The law committee also rejected clauses which would have required industrial hygienists to "fully inform workers and employers about . . . industrial health hazards, and which would have required them not to violate laws and regulations governing the control of industrial hazards." Attempts to debate the code of ethics from the floor of the annual AIHA meeting were quashed by the leadership.[71]

Hiring more industrial hygienists in private industry will do little to improve conditions unless the hygienists are given the power and resources to carry out their work properly. Since they are directly responsible to management rather than workers, they are usually ignored unless there is substantial worker pressure for cleanup. A former employee of the compensation-safety department in the automobile industry found the working situations of industrial hygienists quite limiting:

When I was at Chrysler they had two [industrial hygienists] for the entire corporation. We would have conversations and complain about the workload, and the Industrial Hygienists would say, "Well, there's just no way that we can handle all this. Furthermore, when we go into the plant, the safety man knows we're coming, he opens the windows, he opens the doors, he turns fans on so that the atmosphere doesn't appear unsafe."[72]

Fragmentary evidence from the *Occupational Health Survey of the Chicago Metropolitan Area* suggests that the presence of company-employed industrial hygienists (usually in very large plants) may have little positive effect on the environmental quality within those plants.[73]

Thus despite their achievements in creating the TLV concept, writing and printing the widely acclaimed *Industrial Ventilation Manual* (84,000 total copies sold between 1951 and 1971), and winning new jobs and high salaries for themselves, industrial hygienists have usually felt that as a group they have been denied the public recognition they deserve by forces outside their control. George Clayton, the executive secretary of the American Industrial Hygiene Association (AIHA), expressed this malaise:

Historically we have found if money is given to State governments in relationship to safety and health, even in the general terms of health, that the administrators find ways and means to utilize this money in areas other than industrial hygiene which is devoted specifically to the health of the working man and the control of the environment. . . . Why this is I am not sure. Maybe it is not a very dramatic profession.[74]

Perhaps E. J. Baier, former head of ACGIH and deputy director of NIOSH, most poignantly expressed the feelings of insignificance to which industrial hygienists are prone, saying, "No mother raises her son to be an industrial hygienist!"[75]

Industrial Nursing

With 20,000 practitioners, nursing is the largest health and safety specialty. It has been primarily a women's occupation, with the subordinate status that implies in a male-dominated society. Occupational health nurses have little formal authority at work, though they are the first contact injured workers usually make with the industrial medical system. They work directly under physicians or under "standing orders." Most receive very little training, either in nursing school or afterwards, in the recognition of occupational disease and its prevention, and company medical manuals devote little attention to their tasks.[76] Although their responsibilities for medical examinations, health monitoring, histories, and record-keeping have increased since the passage of OSHA, the law pays little attention to them. Marjorie J. Keller, a registered nurse and an associate professor of nursing, has written that federal training grants are unavailable for occupational health nurses, and that there is little interest within nursing in industrial problems.[77]

Nurses represent the greatest pool of misused talent in the occupational health specialties. If they were given the training and authority to investigate accidents and health conditions they could tremendously increase their contribution to worker health. As it is they already have more intimate contact with the health problems of the workplace than the physician. Such a development, however, would require a complete change from the dominant concept that the industrial nurse is an extension of the company doctor.

Safety Engineering

Most industrial practitioners learn safety work "the hard way [in] the proverbial 'school of hard knocks' " according to the secretary and managing director of the American Society of Safety Engineers (ASSE).[78] Despite the professionalistic trappings of the 11,000-member ASSE, a survey of educational opportunities found that only New York University offers a Ph.D. in industrial safety. In addition to on-the-job experience some training is carried out in short courses by the National Safety Council and at some schools of higher education. Like the organized representatives of industrial medicine and industrial hygiene, the ASSE favored the pro-business bills on occupational safety and health, and displayed a great reluctance to accept the idea that companies should be required to observe government-imposed safety and health standards.[79]

Ambitious safety specialists are concerned about their lack of prestige. The chief of accident prevention of the Imperial Oil Co. of Canada, Hugh M. Douglas, complains that safety is "seldom . . . called into profit planning in the initial stages of any corporate planning. . . . It hasn't been included 'in' or 'out'; it simply hasn't been seriously considered."[80] For most safety engineers concerned about their work's lack of status, the relevant reference group is top corporate management. They like to call themselves "safety managers" rather than "safety engineers" and have founded a group called the National Safety Management Society. A recruiting brochure subtitled "The Director of the Profession" promotes "new concepts and new techniques of [loss control]. . . . The Society expands and promotes the role of safety management as an integral component of total management . . . to improve systems of management toward the purpose of controlling accidental losses, whether they be personnel, property, or financial."[81]

In this National Safety Management Society pamphlet the low estate of safety is alternately blamed on upper management ("any progress must start at the top") and on the safety specialists themselves (who don't "relate to the same objectives as management as a whole"). The proposed solutions to the prestige problem involve a combination of sales technique, dogged persistence, and service, all aimed at convincing top management of the usefulness of safety specialists. All these approaches take the present institutional context for granted, asking for more within the present system. There is no mention of changing that context—of educating workers to demand their rights under the law or asking for more adequate compensation for accident and disease victims.

Similarly, sophisticated books, such as Gilmore's *Accident Prevention and Loss Control* (1970), conceptualize "loss control" as if workers and unions were passive objects of policymaking. The assumption implicit in the work is that it is in management's interest to prevent some work injuries. Occupational diseases are not mentioned as a problem. "Loss control" is the "art of attaining the optimum balance of loss potential, loss probability, and profit."[82]

The trouble with this cost-accounting mentality for the employee is that it does away completely with management's incentive to prevent accidents when compensation payments are low or nonexistent, or cost less than it would cost to prevent injury or disease. The Chrysler Corporation, for example, weighs the costs systematically, and has found it cheaper to pay compensation for most injuries and deaths than to attempt to prevent them:

> Judge Philip Colista described the Chrysler safety program at Eldon Gear and Axle [Detroit] as abominable; that was, I think, typical of the whole corporation. Let me just illustrate it this way. Every year the compensation reps at the various plants were instructed to compute an estimate of the Workmen's Compensation costs for that plant for the coming year and we had to turn those estimates in to the accountants for the corporation. The safety personnel at the plants did the same. They computed their costs ... then it was just a question at the corporation of deciding which is cheaper, to take some injuries, take some deaths, pay some Workmen's Compensation or spend a lot of money and make it safe.[83]

Health, Hygiene, and Safety Under OSHA

In spite of the fact that their professional associations opposed a strong occupational safety and health act, safety engineers, industrial hygienists, and industrial physicians have been among the chief beneficiaries of the OSHA law of 1970. It has created a new demand for their services and increased their prestige within many companies. One safety director stated:

I am not a proponent of Big Government, nor do I think the Federal Government should venture too deeply with regulations and controls into American business, but I must admit that, from my point of view as safety director, this law has benefited my program. For me, at least, it has done more good than harm.[84]

Bill Miller, manager of corporate safety and workers' compensation administration for Goodyear Tire and Rubber Company, affirmed this:

If you're not upbeat about OSHA when you talk with top management and employees in your company you are not doing your job. If you can't point out some advantages of safety to top management and employees you don't belong in the safety director's chair. The safety director must show management that safety is just as much a part of production as engineering, quality control, personnel, and company function. He must sell his program. OSHA has made safety a necessary part of production and has been a tremendous sales tool for the safety director.[85]

In recent years the American Society of Safety Engineers (ASSE) has moved to "professionalize" the specialty. It decided in 1968 to sidestep separate state engineering certification and set up a "single, national certifying board for the safety profession," mostly because it would have been impossible to extract legal recognition from state boards of engineers. In 1970 the Board of Certified Safety Professionals was created and there are now 3,500 "certified safety professionals." To become one the candidate must have a bachelor's degree, preferably (but not necessarily) in science or engineering; five or more years of "professional safety experience" with at least two years in "responsible charge" of a safety program; and the candidate must take an examination made up by the board. Two years of qualifying experience would be equal to one year of university training. The president of the Board of Certified Safety Professionals wants states to license safety engineers by having them recognize the board's certification. The president of the American Society of Safety Engineers has made it clear that the organization's first goal is to require OSHA to hire "certified safety professionals" to key federal positions. Such a step has already been recommended for state positions by the National Advisory Committee for Occupational Safety and Health. The justification, of course, is "the need to protect the public from potential harm at the hands of the unqualified."[86]

The process of acquiring legal control of training and entry to a work specialty is a political one where government is induced to grant the monopoly to a particular group. The important thing to remember about professional societies is that their basic purpose is the defense of their members' immediate interests in prestige and income. One commentator put it this way: "Indeed, so far as terms of work go, professions differ from trade unions only in their sanctimoniousness."[87]

The effect of the ASSE certification standards, if accepted by government, will be virtually to exclude from government inspectors' positions union safety people from the shop floor, since they generally haven't finished college and certainly haven't worked in "responsible charge" of a safety program. A uniformly management orientation would be assured for federal and state inspection staffs, a guaranteed supply of government jobs created for "certified safety professionals," and continued laxity of enforcement assured by safety specialists who were trained and indoctrinated with the corporate viewpoint and would inspect factories where their former colleagues worked. The educational requirements specified by the ASSE amount to little more than a college degree and on-the-job training. Without a role for workers and unions, "professionalization" will be a perfect arrangement for perpetuating the existing neglect of people's rights on the job. The only real novelty, the requirement for the passing of a written test, is good if the test is good, but mostly irrelevant to task performance on the job where all other incentives within the system remain the same. So long as workers have little to say about how their jobs are organized and so long as they continue to bear almost all of the costs of occupational disease and injury with their bodies and their lives, the safety engineer's job will be an unimportant one, no matter how many tests are passed.[88]

If the ASSE takes over certification of government health and safety inspectors, moreover, it will be very difficult to dislodge in the future, since the compensation-safety apparatus will defend the right of the ASSE to define who shall work in safety inspection jobs. In addition, all the "board certified" safety specialists will have newly valuable perquisites to defend, which would include the best jobs in the government safety and health apparatus. Unions and workers should certainly oppose any attempts to make the ASSE the judge of who is qualified to protect the health of people on the job.

Inside the Corporation: A Short Glimpse at Standard Oil of Indiana and the Budd Co. of Philadelphia

Evidence about the functioning of the medical department of Standard Oil of Indiana (of which Amoco is a subsidiary) shows how a sophisticated corporate medical program defends the financial interests of the company. In 1973 the organizational chart of the company's Medical and Environmental Health Services showed that they employed forty-five health and safety specialists, including eight physicians, four industrial hygienists, two toxicologists, one industrial hygiene chemist, seven safety engineers, one safety analyst, and twenty-two nurses. The director of this substantial program was a physician named Peter Wolkonsky, directly responsible to the vice-president for employee and public relations and also a faculty member of the Northwestern University School of Medicine in Chicago.[89]

The staff was responsible for injuries and acute diseases incurred on the job as well as preventing worker exposure to hazards. But testimony collected by the Oil, Chemical, and Atomic Workers (OCAW) shows that Standard's plants around the country have serious occupational health and safety problems. Leo Reidel, president of the OCAW local at Standard's Amoco refinery in Texas City, Texas, believed that nine worker deaths from cancer there were caused by exposure to hydrogen sulfide. John Hocking, a worker at the Texas City refinery, testified at a union conference on industrial hazards that Amoco "had three people killed and several overcome by H_2S gas. . . . It's vented out of the plant and into the atmosphere. We put a chemical in it and the trade name is Mum; it's a perfume. I'd like to know what we're doing, not only to the members, but to the general public outside the fence."[91] Other workers charged that for years the company refused to release to workers information about trade name substances which were causing rashes.[92]

Judging from the company's *Medical Manual* the day-to-day operation of the medical program is directed primarily toward minimizing workers' compensation claims. Physicians handling claims cases must involve a lawyer in treatment decisions at all stages of a potentially compensable injury or disease, even before outside medical advice is sought. The worker-patient is kept in the dark as to the possible nature of medical problems.

The guidelines in the *Medical Manual* speak for themselves:

> Any employee alleging an industrial injury will be asked to fill out, if he has not already done so, a "Report of Accident Form." He will be seen by a company physician, who will, in addition to examining the patient, take a careful history of matters having possible bearing on the complaint; a copy of the medical department report (FORM 53-117) will be sent to the Claims Attorney.
>
> In any questionable case, or in any case in which referral to an outside physician appears to be indicated, the Claims Attorney will be contacted by telephone while the patient is still in the Medical Department. (A typical case would be an alleged back injury where objective findings are lacking.) The Claims Attorney and physician will discuss the case and agree on a course of action. No referrals will be made to an outside physician before contacting the Claims Attorney.
>
> It should be noted that all claims cases are adversary cases, and the comments of company physicians to employees must be guarded as to causation and as to liability for costs. . . .

Because claims cases and absences from work due to real or alleged occupational injuries and illness are a significant cost item for the company, the Medical Department will evaluate such cases critically and frequently as to their ability to return to work, and approve them to do so as soon as possible. Doubtful and disputed cases should be discussed with the Claims Attorney on a regular basis.[93]

The Problem of Obtaining Information

This example illustrates the obstacles workers face in obtaining accurate information on the hazards to which they are exposed. The medical and environmental departments try to withhold information from workers not only about how much their bodies may have been damaged, but also about dangerous conditions within the plant which could lead to costly demands for a cleanup. This secretive mode was exemplified in a memo by Paul D. Halley, Standard's environmental director, recommending a strategy to medically screen employees in one department of Standard's Amoco Chemical plant in Joliet, Illinois who were known to be exposed to dangerous levels of airborne dust and acetic acid. The problem was to figure out how to check the workers at risk without letting them know they were in danger:

It is suggested that a chest X-ray (14" x 17") of persons exposed to a combination of dust and acetic acid vapors might be in order. It may be possible to obtain such X-rays of the entire plant personnel through the mobile X-ray unit of the Tuberculosis Association or the State Department of Health. This would avoid calling undue attention to any one group of employees. . . .[94]

William Curran, writing in the *New England Journal of Medicine*, holds that physicians rarely have the right to reveal information about a patient except under court order.[95] This would preclude, of course, telephone conversations with attorneys during a medical examination which are routine at Standard of Indiana.

A long-term function of Standard of Indiana's in-plant medical and environmental program is to lobby for weak governmental regulations on health and safety, by using its factories as laboratories and workers as guinea pigs to produce data to support the company's quest for lax standards. On July 22, 1971, for example, a letter was sent from the American Petroleum Institute (API) to members of its Committee on Medicine and Environmental Health, stating that NIOSH would soon begin research on a proposed standard for exposure to heat. It requested "assistance in accumulating pertinent data which are needed to define standards that will be both relevant and feasible" and added, "We would be happy to forward your data [to NIOSH] through API if you wish. If you decide to do so, we could submit the data without specifying your company name." It concluded by saying, "The importance to the industry of this effort does not need elaboration." Environmental director Paul D. Halley (later elected president of the American Industrial Hygiene Association) replied:

> We have done essentially nothing on evaluation of heat exposures in the company. . . . According to Dick Brief, Esso, the proposed ACGIH (will now be OSHA) [standard] will be most difficult to meet in petro. industry. He asks for help to combat the proposals and get HEW [NIOSH] to revise their thinking. They agreed to a delay in standards to July 1972 and further research.[96]

As of September 1978 OSHA still operated without a heat standard.

In dealing with OSHA inspections, privately employed health and safety specialists are required to defend the company's interests by covering up violations as much as possible. The policies of the Budd Company's Trailer Division, which manufactures over-the-road truck components, are typical. Management personnel are instructed to admit OSHA compliance officers to the plant, unless the Plant or Branch Manager "knows that an inspection would probably disclose serious violations." Afterwards they are instructed to play a cat-and-mouse game:

As soon as a compliance officer is known to be on site, all supervisory levels should be alerted and asked to review their respective areas in anticipation of a walk-through inspection. Often many potential violations can be corrected before the inspection of a particular area. This is especially true of the "housekeeping" type of violations—unobstructed aisles, clear floors, grounded wiring, no broken ladders in use or in sight, etc. Endeavor to "stretch" the pre-inspection conference noted in Paragraph A to allow supervisors to present the best possible work areas.

During the tour the Compliance Officer is to be taken where management desires *and no other place* unless so requested. However, the Officer cannot be denied any area which is specifically requested. A representative from management, preferably the Plant or Branch Manager and the Facility Safety Representative and an hourly employee, when requested, should accompany him. Use discretion in deciding where to begin the tour and which areas to tour in what order.

The Compliance Officer is to have all questions answered without elaboration, explanation or evasion. Wherever possible, the Officer is to have either an affirmative or negative answer— unnecessary information is not to be volunteered. Nor should you guess or express an opinion as opposed to a fact. The accompanying member(s) of management should be positive and relate in detail the Trailer Division's progressive safety posture and cognizance of the welfare and safety of its employees.[97]

The New Activists

Building a new health and safety movement would have been impossible without the commitment of hundreds of new activists, often with well-developed technical skills, who have a set of operating assumptions almost diametrically opposed to establishment views. They blame injuries and occupational diseases primarily on the unwillingness of corporations to spend money to design a safe and healthy workplace, and on the constant drive to speed up production. They believe workers should participate in the design and control of production equipment; that progress can be won only by educating and organizing workers and unions to take strong and informed positions on health and safety; and that workers should have the right to walk off unsafe jobs until conditions are corrected.

The new activists come from a variety of backgrounds. Chemist Jeanne Stellman and physician Susan Daum wrote *Work Is Dangerous to Your Health*, the first textbook of work hazards to be written in everyday English.[98] A number of people, including this author, helped found the Chicago Area Committee for Occupational Safety and Health (CACOSH), a coalition of workers, unions, and activists which has led the fight for better working conditions in the Chicago area. CACOSH was founded in January 1972 at a conference co-sponsored by the Medical Committee for Human Rights, the University of Illinois School of Medicine, and a large number of local unions, mostly from the auto, petrochemical, and steel industries. One of CACOSH's first heads was Carl Carlson, a blacksmith and longtime safety chairperson at United Auto Workers Local 6 who has been investigating noise and other hazards at the International Harvester plant since 1959.

CACOSH holds an annual conference with a different major theme each year. It has given dozens of classes for unions on health and safety, and whole courses for the Steel Workers and the Oil, Chemical, and Atomic Workers. The organization has led successful campaigns against the Illinois state OSHA law and for the passage of a law to give compensation for partial hearing loss. It has also worked closely with Dick Marco and Joe Naughton of UAW local 588 in testifying for strong federal noise and power press standards. CACOSH is supported primarily by dues assessed to union locals, and has provided inspiration for similar coalitions in cities around the country—in North Carolina, Philadelphia, Boston, Los Angeles, and Rhode Island, some of which are still thriving.

Local health and safety groups have received excellent support in Washington from the Health Research Group, founded in 1971 by Ralph Nader. Much of the work of HRG focuses on the setting of new health standards for cancer-causing substances, and in watchdogging the OSHA law enforcement apparatus in cooperation with a number of unions, especially the Oil, Chemical, and Atomic Workers. Andrea Hricko, with a master's degree in public health, and Bert Cottine, a lawyer, were the backbone of the group's effort in occupational health for its first three years. In addition to their work with the regulatory agencies, they uncovered a factory of the Rohm and Haas Corporation which knowingly killed dozens of workers by exposing them to the cancer-causing chemical BCME in Philadelphia.[99]

In the area of research on occupational disease, only a few physicians outside of the establishment orbit have been able to fund their research projects on a scale comparable to that of industry. Dr. Selikoff, who became famous in the battle against asbestos disease, is the best known. His Environmental Sciences Laboratory at Mt. Sinai Medical School in New York employs forty-nine scientists and has the ability to respond quickly to new work environment problems as they present themselves. In spite of the funding crisis in medical research, the budget of the Environmental Sciences Laboratory grew from $800,000 to $1,000,000 between 1972 and 1973. The laboratory was awarded a grant of $2.9 million in the autumn of 1973 by HEW's National Institute of Environmental Health Sciences. Most of its studies on occupational health topics are done with the cooperation of unions, but the laboratory also does some work for industry.[100]

Dr. Selikoff is one of the founders and the first president of the Society for Occupational and Environmental Health (SOEH), primarily concerned with promoting research and training in occupational health "without fear or favor."[101] Its founders believe that organizations such as the Industrial Medical Association (now the American Occupational Medical Association) have represented only big business and insurance points of view. By contrast, anyone can be considered for membership in SOEH on the recommendation of two members even if they lack professional degrees. The officers and councillors of the society are mostly university researchers, as well as a few company doctors and trade union representatives. SOEH is the first professional society dealing with

working conditions which has actively solicited union participation. The strategy of the society is to preserve the outward
forms of neutrality between management and labor, in order
to make occupational health a respectable research issue
which the government and foundations can support without
an appearance of partisanship. As such, corporate physicians
are routinely elected to high but secondary positions in the
organization, and financial support has come from corporate
and foundation sources, as well as from dues. Even so, the
Industrial Medical Association's *Journal of Occupational
Medicine* questioned the need for forming SOEH, which
it correctly interpreted as a challenge to its position of leadership in occupational medicine.[102] The professional prestige of
the leaders of the Society for Occupational and Environmental Health has made it the single most important group to
bring the issues of occupational health before the medical-
scientific community and the public at large.

The presentations of the society have covered a wide variety of topics, such as "Current Issues in the Setting of Occupational and Environmental Health Standards," a "Workshop
on Directions for Occupational Health in the United States,"
and a large conference on shipyard health. The sessions at
this important early presentation of SOEH, the "International Shipyard Health Conference," were probably too technical for most workers and union staff people to understand.
The timing of the conference, from Thursday morning
through Saturday morning, made it very difficult for rank-
and-file workers, who do not control their work schedules, to
attend. With a registration fee of $50 ($10 for students), plus
transportation, room, and meals, it is hard to see how many
ordinary workers or safety committee people could afford
such a conference. Trade unionists made up a tenth of the
speakers in the official conference program.[103]

The minutes of the society's "Workshop on Directions for
Occupational Health in the United States" reflect an inability
to decide what directions to recommend. For example, there
was agreement that company physicians were either used by
management or ignored on matters affecting workers' health
on the job. But the solutions proposed during the discussion

seemed contradictory. At one point it was suggested that "some kind of incentive to compliance [with OSHA regulations] should be sought, since fiat alone tended to accentuate resistance and to delay acceptance" of necessary changes. Others argued that "conflict is necessary if advances are to be made." The summary of the meeting concluded that "the views presented and positions taken at this meeting were quite mild; much milder than those taken by some of the participants in other places. . . . In some ways this mildness is disturbing, and may suggest that the SOEH meeting contributed little to improvement of the situation."

SOEH can be expected to push for more research monies in occupational health and safety research, and those members who are not financed by corporate money can be expected to present honest views about the hazards of work. The society has published a pamphlet written in popular language called *Cancer and the Worker*.[104] But there are certain limits to what SOEH can do. Because of its tax-exempt status it cannot engage in overt advocacy for stronger standards. Views which had been considered radical before the passage of OSHA can now receive a respectable hearing and be taken seriously by the establishment media. But given the differing interests of the unionists, company physicians, university researchers, and government administrators who comprise the membership, it is difficult to imagine that the society will be used as a decision-making forum by worker-oriented groups.[105]

A number of projects have sprung up at labor-university centers to deal with occupational health and safety issue. Perhaps the best known is the Labor Occupational Health Program (LOHP) at the Institute of Industrial Relations of the University of California at Berkeley. The Berkeley program has been funded mostly by the Ford Foundation, and has provided a great deal of technical support and training to unions in northern California in occupational health and safety, and has published some very useful reference works on health and safety, notably Andrea Hricko's book-length *Working for Your Life: A Women's Guide to Job Health Hazards* (1977). The staff at LOHP is a very competent amalgam of veterans of the Naderite Health Research Group, the Medical Committee for Human Rights, the International Association of Machinists, and the Oil, Chemical, and Atomic Workers unions. Since it is located inside the Institute of Industrial Relations, an institution which Ford has funded in the past, the project has been unable to carry out freewheeling investigations and lobbying in the style of the Health Research Group or the COSH groups, and it ran into OSHA reluctance to refund its union apprenticeship training programs when it offended the foundry industry with its blunt portrayal of the dangerous conditions in a film entitled *Working Steel*. On the basis of conversations with LOHP staff members and COSH activists, I concluded that the relationship between LOHP and the various COSH groups is one of mutual support.

Though the networks of unions, activists, and researchers have hardly dominated policy in health and safety, they have managed to create a permanent opposition to the compensation-safety establishment on all levels. A few of the new people have secured jobs in the government inspection apparatus, and some government inspectors have participated in groups such as CACOSH and Science for the People, where they have experienced the necessity for strong alliances with labor in order to further their work. The new groups have given invaluable support to people in workplaces and have furnished a new generation of union insurgents with ammunition and programs which will be a permanent part of the labor movement in the United States.

|5|

The Workers and the Unions

In the United States, the struggle for safer work has been carried out primarily through the unions. And it may be that the struggle for better working conditions, with its emphasis on worker rights at the point of production, will help re-kindle the U.S. labor movement. More and more people are realizing that big money is worthless if you are too sick and tired to spend it, and that a good pension becomes null and void if you do not live to enjoy it. Throughout the country workers are demanding to know how the production process affects their health, and for the first time in recent history some of the unions are systematically searching for the answers. Where the struggle has been most intense, in the coal mines and in the chemical and asbestos industries, the bosses have been forced to share a bit of their authority over the productive process.

Most of organized labor's advances began during the 1930s Depression, which drove desperate workers to a series of pitched battles and strikes in defense of their right to earn a decent living. The tight labor market during and after World War II and the great strike wave of 1946–1947 served to cement the position of the industrial unions in the steel, rubber, and automobile industries. The corporate counter-attack relied heavily on legislative constraints to union organizing, such as the Taft-Hartley Act, which abolished the closed shop and encouraged states to pass anti-union "right-to-work" laws. Taft-Hartley also prohibited Communists from holding union office, and helped to create the red scare that drove many progressives from the union movement. As a result, the proportion of workers in unions peaked in 1956 at a quarter of the workforce, and then slowly declined.[1] Although unions have recently had difficulties organizing new shops in manufacturing or the services, a great many government workers have been organized since the late 1960s. Today, in the late 1970s, unions are trying to regain the offensive with a national legislative campaign to change union organizing and recognition rules in their favor.[2]

Unions are strongest in large-scale industries like coal, automobiles, steel, and trucking,[3] where they have organized most of the workers. In industries like these, their members can strike and halt production without much assistance from other unions or the public. Wages are relatively high and the corporations can maintain their profits by raising prices together and by the introduction of labor-saving technology. In other monopolistic industries, such as petrochemicals, the unions are weaker because they control a minority of the work force and because production is so highly automated that supervisory personnel can maintain production during a strike.

In electronics, textiles, clothing, and agriculture the unions have a much harder time surviving. Most enterprises are small and nonunion, the work is labor-intensive, and business is very competitive both at home and abroad. The unions have little control over the labor market, especially in "right-to-work" states where union membership is not compulsory when a shop is organized. Employers in the competitive sector can often hire undocumented foreigners, subject to instantaneous deportation, for as little as five dollars a day, or contract the work abroad for even less. Workers in the lower paying industries are often women or black, Hispanic, or Asian in background, and the better jobs are closed to them by sex and race discrimination. In this sector the unions, where they exist, are usually in a poor position to improve working conditions and wages. Sharp benefit increases can drive individual employers out of business, as unions are well aware. Under such circumstances, organizing a new union is very difficult without interunion solidarity and public support.[4]

During the period between the enactment of state workers' compensation legislation at the start of the twentieth century and the passage of the Coal Mine Health and Safety Act and OSHA, most collective bargaining contracts focused on safety equipment, worker protective devices, and hazard pay, where they dealt with the health and safety issue at all. Where management-union safety committees were created they had little formal authority. A few unions, such as the United Auto Workers, always had the contractual right to strike over safety, and many unions, such as the International Longshoremen's and Warehousemen's Union (ILWU) and the International Brotherhood of Electrical Workers (IBEW) have traditionally given workers the right to refuse dangerous work. Today contract language is much more likely to emphasize worker rights to full information, to medical exams, and to protection against firing or other discrimination for raising the health and safety issue. Now the boss has a duty to provide a safe and healthy workplace.[5]

An Upjohn Institute survey conducted before OSHA came into effect showed that good health and safety practices and contingency protections (such as workers' compensation) were top priorities among unionized rank-and-file auto and steel workers. But most striking was the similarity of opinion between top union leadership and upper management. Both thought that money rather than working conditions deserved the most attention, an almost exact reversal of blue-collar attitudes. A national sample survey of employed persons corroborated the importance workers (particularly those who work with their hands) give to health and safety conditions, and confirmed the "same reversal of worker priorities by the union hierarchy."[6] So it is no wonder that solid programs to prevent industrial casualties before OSHA were almost always the result of visionary local leadership or mass uprisings, without the support of national union leadership. Only the miners' unions at different times in their histories have been an exception.

However, while state-level lobbying on industrial inspection and workers' compensation was left largely to insurance and business interests before OSHA, improvements in workers' compensation benefits were invariably the result of union pressure. Where unions were weak, particularly in the South, payments were low.[7] Only the passage of the OSHA law, under heavy pressure from the unions, finally generated the mass expectations from below which have made it impossible for U.S. labor organizations to ignore the problems of occupational health and safety.[8]

The situation of the increasing majority of nonunion workers is much weaker than that of the organized. Unless a catastrophe occurs, their safety is ignored. Their only options for a dangerous situation are either to suffer in silence or quit the job and go hungry. Those who protest or file OSHA complaints run the risk of firing even though OSHA regulations forbid this practice. The exercise of workers' rights is difficult even with the aid of the union structure. Nonunion workers should not expect much from OSHA, unless they use it as part of an organizing strategy.

Oil, Chemical, and Atomic Workers (OCAW)

Working in a refinery is normally a dangerous, filthy job. A corporate history of Standard Oil of Indiana described one operation in the following terms:

> The noxious sulphur dioxide produced in roasting the copper sulphide, and the fine floating dust made the millhouse an inferno. It was the dirtiest and most unhealthy work in the refinery. . . . The work was done mostly by foreigners. Workmen had to wear respirators, and most men could stand only a few days of it. If they stayed too long, the fumes ate the skin off their faces and made their eyes bloodshot.
>
> One of the great curiosities about the refinery was an old white horse that hauled the copper sulphide to the dump. Its hair had turned green.[9]

An early strike in the oil industry over working conditions occurred on July 11, 1915 in Bayonne, New Jersey. The still cleaners were protesting, among other things, that the company made them enter the stills before they had cooled. "The still cleaner's job was perilous. Dressed in padded suits for protection from heat as high as 250° F, they chipped the coke that remained on chamber surfaces after final products had been extracted. Paid $2 to $2.40 a day, stillmen probably developed black lung, the disease of the coal miner, from constant exposure to coke dust." Eventually Standard Oil broke the strike with the help of the courts and police and a battalion of scabs, but only after killing nine workers and injuring at least fifty.[10] Health issues at work triggered early unionization at the Pasadena, Texas, Shell refinery, and in recent years the 180,000-member OCAW has become one of the leaders in the campaign for better working conditions.[11]

The OCAW believes very strongly in the importance of the education and training of its members about job health and safety. Prior to the passage of the OSHA law, the union sponsored weekend conferences around the country accompanied by knowledgeable scientists so that "workers themselves could tell the story of the conditions they encounter daily." Ray Davidson, editor of the union's newspaper, wrote an exposé of conditions in the oil and chemical industries, *Peril on the Job* (1970), which has yet to receive the attention it deserves.[12]

The OCAW's national program, under the leadership of Tony Mazzocchi, now an international vice-president, has consciously worked to build coalitions with the scientific, environmental, and medical communities. In the fall of 1971, the now-defunct Scientists Committee for Occupational Health ran a pioneer course sponsored by District 8 to teach workers how to understand and deal with occupational health questions. The teachers wrote a handbook in lay language which has become a model for much subsequent education around the issue. The Occupational Health Project of the Medical Committee for Human Rights (MCHR), for example, borrowed heavily from this manual in their long pamphlet, *Health Hazards in the Workplace* (1972). Doctors Jeanne Stellman and Susan Daum published a revised version called *Work Is Dangerous To Your Health* (1973), which has become the most popular reference work in the field, having sold tens of thousands of copies around the world. Stellman later became the assistant in occupational health to OCAW's President Grospirin. The union has also worked closely with grassroots coalitions such as the Chicago Area Committee for Occupational Safety and Health (CACOSH) and the Philadelphia Area Project for Occupational Safety and Health (PhilaPOSH). On the national level, program director Mazzocchi has been adept at mobilizing the mass media to create a new climate of opinion within his union and throughout the United States. Paul Brodeur's *New Yorker* series on the asbestos poisoning of OCAW workers at a Pittsburgh-Corning plant in Tyler, Texas, was first made possible by Mazzocchi's careful cultivation of the media.[13] Only Rachel Carson's "Silent Spring" series has ever generated more reader letters to the *New Yorker*.

In January 1973, four thousand Shell workers in the OCAW walked off their jobs, in a strike triggered primarily by work environmental issues. Shell, almost alone among major oil companies, had refused to (1) allow independent scientific surveys of working conditions; (2) pay for appropriate medical examinations; or (3) provide the union annually with all available information on the sickness and death rates of its employees. When the strike began, OCAW appealed to outside environmental, professional, and public interest groups for support: consumers were urged to boycott Shell products and to turn in their credit cards and demonstrations were organized in several cities to publicize the health issues and increase support for the workers. Many people around the country first became aware of the working conditions issue as a result of the strike. Although production was slowed, it was never completely shut down, due to the inability of the union to organize supervisory personnel and the high degree of automation in the industry. After four and one-half months on strike, the workers went back, having won only the right to information on sickness and death rates. The right to hire independent health consultants at company expense was defeated. But OCAW leaders regard the fact that the strike occurred at all as a forward step for the union and the occupational health movement.[14]

The OCAW attracted the public eye again when Karen Silkwood, a nuclear laboratory technician at a Kerr-McGee plutonium plant near Oklahoma City, was killed under mysterious circumstances. Silkwood had been fascinated by nuclear power since her high school days. She walked the picket line when the union struck Kerr-McGee for a bitter ten weeks in 1972. After the strike, dismayed by the company's lackadaisical attitude toward safety and upset by her exposure to plutonium and the company's refusal to buy her a respirator which fit, Silkwood became a union activist. She was elected to one of the three seats on the Local 5-283 steering committee, and became its chief force for better working conditions.

On November 13, 1974, after taking care of some union business, she left for a meeting with Steve Wodka, of OCAW's Washington staff, and David Burnham of the *New York Times*, who was investigating the dangers of the plutonium industry. Fellow workers remember that she carried a large manila envelope filled with papers under her arm. But she never made it. She died on the way when her automobile ran off the road and smashed into a culvert. By the time Burnham and Wodka tracked down her car at a local garage, the big envelope was gone.

The Oklahoma State Highway Patrol theorized that she had fallen asleep at the wheel, perhaps from an overdose of Quaaludes. Wodka and many others believe she was murdered. Accident investigators hired by the union concluded that Silkwood's car had been bumped from behind by another vehicle. Superiors of the FBI agent in charge of determining whether a crime had been committed discouraged him from speculation about possible suspects or motives in the case. The agent was specifically instructed not to investigate possible plutonium smuggling from the

Kerr-McGee plant. So far, all attempts to trigger a full-fledged congressional investigation of the case have been stymied, and the Justice Department has dropped the case.

Despite the official cover-up, Karen Silkwood has not been forgotten. The National Organization for Women (NOW) proclaimed the first anniversary of her death Karen Silkwood Memorial Day, and many consider her a martyr of labor and women's causes, and even of the struggle against nuclear power. Her parents, at first pleased when the FBI entered the case, now believe that "the FBI has let us down. . . . We taught Karen to speak up when she saw something that was wrong," her father told a candlelight vigil in her memory. "Maybe that's why she's not here tonight. But her death will not have been in vain if you all speak up now and don't let her killers get away."[15]

In recent contract talks the OCAW has attempted to secure up to five days' annual training for their health and safety committee members, access to the identity and hazards of all (especially trade name) substances used in their workplaces, and full pay for safety and health activity, including walkaround time during government inspections. The union has also attempted to extract the maximum concessions from already existing contracts. Local 3-562 at the Ciba-Geigy Corporation in McIntosh, Alabama, won a significant victory when an arbitrator ruled that the local had the right to the generic (chemical) names of all substances used in the plant.[16]

Yet in many ways the OCAW operation is still struggling. With two full-time people in Denver and two part-timers in Washington, it has to fight hundreds of full-time company specialists throughout the United States and Canada and contend with the regulations in dozens of governmental jurisdictions.

The OCAW is also using the health and safety issue as an issue to help organize new workers. A couple of its health and safety staffers have attended the union's school for organizers, and plans are now underway to set up an organizers' school in California in which health and safety will be a major part of the program. Despite internal opposition, the leadership's aggressive stance on health and safety has continued, with ever greater backing from the rank-and-file as they come to understand the issues. In the petrochemical industry, where most of the workers remain unorganized, where speedup is always present and new jobs scarce, public support has been essential in order to continue. And in the process, the OCAW has helped change the whole climate of public opinion in the United States about the issue of working conditions.

United Auto Workers (UAW)

The UAW is the second largest union in the United States, with 1.5 million members. Wages average over $7 per hour, and those with enough seniority can live adequately if not luxuriously. The most progressive large union in the country, the UAW has also won important benefits in the areas of medical and dental care and pensions. These are particularly important in the United States, with its relatively low Social Security retirement benefits and "free enterprise" health care.

Since the union's founding in the 1930s, health and safety has always been important for the union. National contracts with General Motors, Ford, and Chrysler have always permitted local strikes over health and safety, so long as they were sanctioned by Solidarity House, the union's headquarters. In the early 1950s the UAW was one of the first to hire full-time health and safety staffers, and by the passage of OSHA the total number of health and safety staffers had climbed to four.

The UAW lobbied hard for the passage of the OSHA law of 1970, but the initial reaction of Solidarity House to the law was cool and defensive. The union sent its locals a lengthy and legalistic explanation of the OSHA law prepared by a business consulting group with a UAW cover tacked on. This pamphlet contained an administrative letter which called for a centralized approach to the problem. Local unions were encouraged to take their health and safety complaints to their staff representative, who would then decide whether an OSHA inspection was warranted. Almost everybody, however, ignored the letter, and some UAW members refused to believe it had ever been written.[17]

Some UAW educational directors acted to bring in viewpoints and resources from outside the union. In the Chicago area the Region Four director and some of the UAW locals cosponsored a conference with the Medical Committee for Human Rights and the University of Illinois School of Medicine, which led to the formation of the Chicago Area Committee for Occupational Safety and Health (CACOSH). Region Four has let CACOSH use its offices for meetings, and has helped out the organization with a monthly retainer. Ed Graham and Carl Carlson of UAW Local 6 were CACOSH's first and second chairpersons. Similar cooperative programs have begun to take shape with MassCOSH in New England and with PhilaPOSH in the Philadelphia area, prodded by rank-and-file leaders such as Carlson and Jim Moran. Such activists, though not common, exist throughout the union, and have joined with allies on the permanent staff to create a very informal health and safety caucus throughout the union. Some of the health and safety staffers hired at Solidarity House have worked with the COSH groups. Frank Wallick, editor of the UAW's staff newsletter, wrote a book called *The American Worker: An Endangered Species* (1972) which was widely read within the union. By 1973 the union was ready to take a major step in its collective bargaining demands with the Big Three automobile manufacturers.

Contract negotiations between the UAW and the Big Three take place every three years. The union decides what its demands are and throughout the spring and summer of the contract year there is speculation in the daily and trade press about what the demands will be, how much they will cost, and how far the companies will give in. In a ritualized war dance, the industrial and labor giants threaten and cajole, looking for weak spots. One company is usually singled out by the union as the target, and the settlement in those negotiations determines the pattern for the other two companies.

In 1970 General Motors had been the target of a two-month strike which bankrupted the union, so it was clear that either Chrysler or Ford would be the first object of the union's offensive in 1973.

Despite the Nixon administration's wage and price freeze, money was not the big issue in 1973. Cost-of-living and productivity increases had kept the auto workers ahead of inflation, and the plants were running overtime, making lots of cars and lots of money. For the workers the salient issues were the elimination of mandatory overtime and "30-and-out"—the right of workers to retire with a full pension no matter what their age after thirty years in the plants.[18]

But safety and health was certainly a close third for workers and the union. A production workers' conference held in February 1973 showed a tremendous reservoir of discontent over the issue. This session, partially orchestrated by sympathetic Solidarity House staffers, uncovered wide dissatisfaction with OSHA enforcement and the performance of the UAW health and safety staff. Melvin Glasser, head of the union's Social Security Department, stated in a press interview that the union wanted its own members to be trained to handle hazards on the job.[19]

The determination to press for strong health and safety language in the 1973 contract was sealed by a series of wildcat strikes in July and August at some of the antiquated Chrysler plants in the Detroit area. The immediate grievances concerned working conditions or racial harassment of black workers by white supervisors. On July 25 two young black assembly line workers halted production at Chrysler's Jefferson Avenue plant when they locked themselves into the electrical control cage of one of the assembly lines and shut off the power. They were immediately protected by hundreds of sympathetic workers until the company agreed to fire the supervisor who had offended black workers with his racist remarks.[20] A couple of weeks later a local UAW president barged into the top-level negotiations between Chrysler and the UAW and demanded that conditions be rectified at the Mack Avenue stamping plant. He said that the workers were threatening to strike, and that he couldn't stop them.

And the Mack Avenue plant did go out on a wildcat a few weeks later, led by a worker from the Workers Action Movement, an organization associated with the Progressive Labor Party. On August 7, 1500 workers, both black and white, walked out of Chrysler's Detroit forge plant. The members of the local, which did not have a militant reputation, were angered at two recent preventable accidents in which some workers lost their fingers. They complained bitterly about the filthy conditions at the plant, which had gone uncorrected for years, despite reported protests. Douglas Fraser, vice-president of UAW's Chrysler Division (and now UAW president) was surprised by the Detroit forge walkout: "This has caught me absolutely cold. I had no hint of any difficulties at Local 41. Why didn't our people know?"[21]

The workers at Detroit forge refused to obey a court order by a local judge which prohibited further picketing. In response, Vice-President Fraser and other members of the UAW's bargaining team recessed national contract negotiations with Chrysler and toured most of the company's twenty plants in the Detroit area. A report they issued threatened more strikes against Chrysler if the company refused to clean up the hazards.

As a result, Chrysler became the target company in the 1973 negotiations, with health and safety at the top of the list, and the settlements with Ford and General Motors followed the Chrysler pattern. The 1973 agreements created management-union safety committees at the national and local levels. Most important, the UAW won the right to appoint full-time health and safety representatives in plants with at least a thousand workers, with wages to be paid by the companies. Securing pay for these three hundred local safety representatives was particularly important, since employers are not obliged under OSHA regulations to pay for "lost-time" spent on safety. The UAW's national safety and health staff was augmented by younger and better trained personnel with a history of grassroots activism in health and safety with unions and COSH groups. Both national and local safety representatives have the right to inspect any factory at any time, and both the company and the union are required to train the newly appointed local safety representatives. And local strikes can still be authorized over health and safety conditions.[22]

Many of the 1976 contractual provisions concern the union's right to information. The management-union safety committees in the Big Three now must be informed by the company of the chemical names and hazards of potentially harmful substances used in the plant. Workers are supposed to be told of the results of medical tests that the company performs on them, and the union safety committee must be sent the data from in-plant air quality tests.[23]

Most of the full-time safety committee people believe that they have had a positive impact on working conditions. Ed Jackson, alternate safety representative at UAW Local 560 (a Ford plant in Milpitas, California), noted a "marked improvement" since the 1973 agreement. He particularly praised a "fantastic" two-week class in health and safety at the UAW's Black Lake educational camp. Both he and Bob Scott, safety head at Local 1364 (a GM plant in Fremont, California), were enthusiastic about the UAW's health and safety staff in Detroit. Jackson's main complaint was that the local union's safety rep lacked enforcement authority: "If we get an immediate life or death situation we can shut the machines down. But if we can't prove someone is just about to get hurt we get overridden." According to Jackson the union has to exercise constant vigilance or the company will push back any gains, as it did after a 1972 mid-contract strike over working conditions.[24]

UAW Local 588 at the Ford stamping plant in South Chicago has waged a long fight over noise and power press safety. After an intensive educational effort, the bargaining committee led the four thousand-member local out on strike. The company fired the entire committee from their jobs, hoping to deal with a less militant group of workers, but the unity in the local made that approach impossible. President Dick Marco and Recording Secretary Joe Naughton later won an arbitration decision restoring their jobs. In June 1974 and October 1976, Local 588 again struck over health and safety, this time with the full support of the national

UAW staff. In 1976 they won a twelve-point program for the installation of noise control devices according to a schedule, both full-time and standby maintenance workers to keep the machines operating safely, and improved power press rules almost equivalent to a "no-hands-in-the-die" standard. Much of the technical support for the efforts at Local 588 have come from CACOSH's health-technical committee. Dick Marco is a former chairperson of CACOSH, and Naughton a former treasurer of the organization.[25] In plants all over the country the UAW is fighting the companies over the issue of noise. Official policy is to encourage engineering controls rather than the use of personal protective equipment such as ear muffs and ear plugs, and the UAW's newspaper, *Solidarity*, which goes to over one and one-half million UAW families, has written vivid accounts of the problem and how some locals have struggled to cope with it. The Social Security Department of the union has published a very useful fifty-page technical manual on noise, which is probably the most widespread occupational health problem faced by UAW workers.[26]

Dealing with health and safety has been much more difficult in the smaller shops, a problem not unique to the UAW. The health and safety staff in Detroit is too small to service the many demands made on it for technical advice. The hope is that the three hundred full-time health and safety reps at the large plants will be able to help their brothers and sisters in smaller shops. Where possible these groups have received some aid from the COSH groups as well as from the national union. With the gains in the last five years, the groundwork has been laid for a long-term struggle over working conditions.

United Rubber Workers (URW)

The 250,000-member United Rubber Workers union has a history of concern over occupational health and safety in the notoriously dirty rubber industry. Although the URW has not received the same national attention as the OCAW, it has been able to accomplish more for its members through collective bargaining, because of its ability to shut down the major tire companies.

The occupational health issue first attracted the attention of the leadership in 1961, when a strange disease felled many employees at a giant Uniroyal plant in Eau Claire, Wisconsin. At the peak of the outbreak, 312 of the plant's 2,225 workers were out sick with respiratory complaints. Some stayed out a day or two, and others never returned to work. Because of the high number affected in certain departments, the union reasoned that a new vulcanizing process had caused the disease. Uniroyal, however, denied any occupational link to the outbreak, and the company was supported by local doctors who diagnosed the problems as head colds and pneumonia. After continued union pressure the company removed the new bonding agent and hired Dr. John Rankin, head of the department of preventive medicine at the University of Wisconsin, to study the situation. He finally concluded that the disease was work-connected, though he was unable to identify a specific causal agent. Despite Rankin's findings, Uniroyal continued to contest workers' compensation claims based on exposure to the bonding agent. So John Barsamian, the local president, called a mass meeting of the members to discuss shutting down the plant. Barsamian reported, "While I was speaking to the men, a note arrived from the company management agreeing to pay workmen's compensation benefits. . . ."

Altogether, 135 workers finally received benefits for upper respiratory inflammations and 103 of these returned to work without loss of seniority. Peter Bommarito, president of the Rubberworkers, participated in three series of negotiations between 1965 and 1970. These settlements won the permanently disabled $1 million in back pay, a company disability pension, Social Security makeup benefits, and permanent payments from workers' compensation for the thirty-two who would never work again. The workers had to push for every gain they made. According to Bommarito:

This dramatically proved again that in unity there is strength to get things done for people. No amount of money can bring back health to these workers, but at least the award will permit them to live the rest of their lives with some degree of dignity.[27]

Since the Uniroyal episode the URW has become a national leader in dealing with health problems on the job. Before OSHA, on the advice of Dr. Thomas F. Mancuso, the URW signed contracts with some of the major rubber companies which committed them to pay one-half cent (now one cent) per worker-hour into a fund to study occupational disease in the rubber industry. These studies are presently conducted at the Harvard and North Carolina schools of public health. The schools communicate their findings to a labor-management Occupational Health Committee at the national level, where the union is represented by industrial hygienist Louis S. Beliczky. This committee makes recommendations based on the findings. If a situation which is "immediately threatening" to life or limb is discovered at a plant, it must be quickly communicated to union and management at the local and national levels.[28]

The URW pushed very hard for a strong vinyl chloride standard in the United States. Its experience in this effort, presented by URW President Bommarito and industrial hygienist Louis S. Beliczky, figured prominently in the International Conference on Occupational Health in Geneva in 1974. Organized by the International Federation of Chemical Workers, this conference brought together for the first time unionists and scientists from the United States, Western Europe, and Japan for practical discussions on how to protect workers' health on the job.[29]

Teamsters

The Teamsters is the largest union in the country, with well over two million members. The most numerous membership category is truck drivers, but almost every kind of worker can be found among its members. For many years the Teamsters union had the reputation of winning high salaries for its members, particularly the truck drivers, but the union has always been known for its close ties with labor racketeers. At present, the union's hierarchy is under serious attack from the rank-and-file, the press, and law enforcement agencies.

Top Teamster officials live royally on the members' dues. Five jet aircraft and two turboprops worth over $13 million and costing over $2 million annually fly the top leadership on its missions; Frank E. Fitzsimmons, general president, receives an annual salary of $156,000 and a host of other benefits; and Harold Friedman, president of Baker's Local 19, a Cleveland local associated with the Teamsters, received the astonishing total of $352,330 from his various union jobs in 1976.

Compared to the leaders' opulent lifestyles, the Teamsters' efforts in health and safety are decidedly meager. In late 1973 the union created a Department of Safety and Health in Washington, D.C. Its director receives $30,000 per year on top of his salary as a local president in North Carolina. A part-time assistant and secretary complete the office. Each of the four regional conferences of the Teamsters has a safety director, but their duties are unclear. Aside from the newsletter, *Shield*, published by the national safety office, the union does not produce any educational publications by itself, but it has reproduced publications of the Bureau of Motor Carrier Safety for the membership. Each Teamster local is supposed to have a safety director, appointed on the recommendation of the local's leaders. According to the Teamsters' western safety and health coordinator, there is no internal union training of local safety officers, though they are encouraged to attend training sessions at junior colleges and other public institutions.[30]

However, the many abuses of the Teamster leadership have spawned a variety of reform movements organized around rank-and-file issues. A group called PROD, which began as a single-issue organization concerned with health and safety, now claims a membership of over four thousand, and has begun to challenge the existing structure at all points.[31]

To its credit, the Teamsters was the first union to create a national grievance system for health and safety. Though little use has been made of the system, it does provide the rank-and-file with a pressure point on health and safety. Article 16 of the Master Freight Agreement (for long-haul drivers) gives strong guarantees of the drivers' right to safe working conditions, but it is almost universally unenforced in practice. One provision of Article 16 forbids drivers from being required to disobey any federal or state safety laws or regulations. When a driver named Jim Banyard was fired for refusing to pull an overloaded trailer, the Teamsters declined to support him against his employer through the grievance procedures. He got his job back only when PROD sued the National Labor Relations Board (NLRB) on his behalf.[32]

But the effort of Teamsters in health and safety has not been limited to the meager activities of the national office. Many locals have their own programs and their own activists who carry them out, with or without Teamster sanction. Teamsters' members work in a variety of jobs—less than half of them are long-haul truckers—and they have a range of health and safety problems that match the diversity of the work they do.

Local 688 in St. Louis, whose members are concentrated in warehousing and small industry, began investigating conditions at some of its shops in the late 1960s, after workers at a giant smelter complained of lead poisoning and other hazards. In 1971, at the urging of staffer Mike Ryan, chief shop steward Art Button, and this author, the local began a systematic investigation of the hazards at the Crane Corporation factory. Crane produces steel pipe for oil pipelines. The workers take slabs of steel, heat them red hot and bend them into half-cylinders, which are welded together lengthwise to form the pipe. This pipe is tested for defects by the use of surface penetrants and x-rays. Every step in the production process is potentially hazardous.

A program (later written up as "A Job Health and Safety Program on a Limited Budget") was created to deal with the situation at Crane. It presumed that workers, who are traditionally ignored, know a great deal about the dangers of their jobs; that union locals cannot be expected to hire full-time specialists in health and safety; and that the power to require healthy working conditions can come only through the self-education and organization of workers under the leadership of their unions. To deal with the problems in a particular shop, workers were urged to:

1. list problems by department with suggestions for correction;
2. list trade name substances and find out their chemical composition and potential hazards;
3. assess the seriousness of each problem identified in the plant, perhaps by polling the membership;
4. make up a looseleaf diagnostic-educational notebook of the hazards in the plant so that progress in dealing with them can be charted;
5. hold educational meetings and write educational materials based on the notebook;
6. estimate the cost of hazard correction;
7. create a permanent safety committee and write the health and safety demands into the contract if necessary.

The pamphlet contended that government inspections "should be used only when the problems in the shop have been well-studied and understood by the workers. Government inspections are almost useless where they occur outside the context of an on-going union program."[33]

Chief steward Art Button took the results of the study carried out at Crane and wrote a twenty-three-page analysis of the situation which included a list of demands to be implemented. He also triggered an intensive inspection by the National Institute of Occupational Safety and Health (NIOSH). Within seven months of the completion of his analysis, most of the modifications had been carried out, and the method used at Crane has become a permanent part of the local's educational program.

United Mine Workers (UMW)

The Workingmen's Benevolent Association in the Pennsylvania coalfields was a forerunner of the United Mine Workers. It got its first great influx of members in 1869, when all 179 miners in the Avondale Mine in Luzerne County were burned to death because the mineowners had refused to build an escape exit. Two days later, as the bodies were dragged out, Irish John Siney, president of the association, spoke to the people who had gathered to pay their last respects and to weep: "Men, if you must die with your boots on, die for your families, your homes, your country, but do not consent to die like rats in a trap for those who have no more interest in you than in the pick you dig with."[34] Thousands joined the association that day, although it was later smashed in a bloody guerilla war with the mineowners.

In its founding preamble in 1890 the United Mine Workers dedicated itself "to reduce to the lowest possible minimum the awful catastrophes which have been sweeping our fellow craftsmen to untimely graves by the thousands by securing legislation looking for the most perfect system of ventilation, drainage, etc." and "to enforce existing laws, and where none exist, enact and enforce them, calling for a plentiful supply of suitable timber for supporting the roof, pillars, etc., and to have all working places rendered as free from water and impure air and poisonous gases as possible."[35] But the laws were ignored even when they were passed, and thousands of coal miners were killed every year. One explosion in Monongah, West Virginia killed 361 miners in 1907. More recently Elmer Yocum, a West Virginia coalminer, charged that most state mine inspectors had worked as company supervisors and were still beholden to "King Coal":

> I don't think any man in here is naive enough to believe that the inspector is going to shut down a mine for a safety hazard and risk the censure of his fellow inspectors, lose his dividends, besides the repercussions he might possibly get from the State department of mines. . . . I have gone into a mine immediately following a State inspector and found violation after violation, and I am not a trained inspector.[36]

As miners fought for their lives on the job, they had to face constant attempts by the owners to bust the union. During the 1920s the UMW was almost completely driven from West Virginia. With the election of Franklin Roosevelt to the presidency and the passage of the National Industrial Recovery Act (NRA) in 1933, UMW organizers fanned out across the coal regions. Membership rose from one hundred thousand to four hundred thousand a year, and UMW President John L. Lewis became the best-known figure in labor. But the UMW's power was soon weakened. Lewis agreed to support mechanization, which cut rapidly into the workforce after World War II, and the switch to oil and gas sharply reduced the demand for coal. Furthermore, the union became as tied to high productivity as the owners, since royalties to pension and welfare funds were determined by a tax on each ton of coal rather than according to each hour worked. By 1972 the workforce was only a third what it had been in the 1940s, although the same amount of coal was being produced. Meanwhile, the profits from coal continued to leave the region.[37]

To get recognition for black lung as an occupational disease, the coalminers and their allies had to fight all the recognized scientific wisdom on the subject. State and federal compensation authorities denied for decades that coal dust was a hazard, though it was recognized as a compensable disease in Britain in 1943. Although it purports to cover all phases of industrial safety and health, the National Safety Council's most ambitious publication, the 1654-page *Accident Prevention Manual for Industrial Operations* (1969), contains not a single hint that coal dust by itself is responsible for the production of lung disease. The Council's 990-page *Fundamentals of Industrial Hygiene* (1971) uses the code word *anthracosilicosis* to describe black lung. The implication is that coal dust causes disease only when mixed with silica. The only passages of the book dealing with the effects of coal dust read as follows:

Anthracosilicosis, a complex form of pneumoconiosis, is a chronic disease caused by breathing air containing dust that has free silica as one of its components and that is generated in the various processes involved in mining and preparing anthracite (hard) coal, and to a lesser degree, bituminous coal.

The disease is characterized anatomically by generalized fibrotic changes throughout both lungs and by the presence of excessive amounts of carbonaceous and siliceous material.[38]

The December 1968 edition of the *Industrial Hygiene Digest* (published by the Industrial Health Foundation) wrote that "coal dust is not the cause for breathlessness in coal miners when it occurs and . . . attempts to relate the two are sociologically rather than scientifically based. . . ." The article went on to suggest that the consequences of assuming a connection between lung disease and coal dust "would be that thousands of coal miners may become victims of iatrogenic [doctor-caused] disease."[39]

The UMW has always paid lip service to safety and health. For decades the *United Mine Workers Journal* has carried memorial articles about the great coal disasters, and occasional educational material on safety and on black lung disease. In 1946 the union's Anthracite Health and Welfare Fund, in an agreement with hard coal operators, provided treatment for black lung victims at the Jefferson Medical College in Philadelphia. For years the union went along with the companies' myth that bituminous (soft) coalminers were not subject to lung disease. Union pressure made black lung a compensable disease in the United States for the first time in Pennsylvania in 1965.[40] But when it came to organizing an active program to protect miners from accidents, coal dust, and other hazards, the national union staff was inactive or hostile. The safety department's concepts were borrowed directly from the coal companies, as though management held the miners' welfare foremost. If the safety department did more than attend industry-sponsored activities, it was not made public. An account of one such meeting published in the UMW *Journal* reads like a National Safety Council press release:

Representatives of the UMWA attended meetings of the National Safety Council which were held here Oct. 27–Nov. 2 in connection with the 60th National Safety Congress.

The UMWA is represented on the Executive Committee of the Coal Mining Section by VP L. J. Pnakovich, Safety Director Kenneth F. Wells, and Rex Lauck, Assistant Editor, the *Journal.* . . .

The Coal Mining Section on October 30 heard papers on roof control: "Resin Anchored Roof Bolts—A New Approach to Rock Strata Control in Bedded Formations," by William Beerbower, of E.I. Dupont and Co., and "Accomplishments in Design of Roof Bolt Materials and Installation Apparatus," by a panel made up of John Parsinger of Acme Machinery Company, Robert Fletcher of M. Fletcher and Company, John B. Long of Marmon Research Corporation and Harley Pyles of the MFC Corporation. . . .

Incoming Chairman of the Coal Mining Section is Alex Keleman, Safety Director of the Coal Division, Armco Steel Corp. He succeeds Edward Onuscheck, Safety Director, Rochester and Pittsburgh Coal Co.[41]

The movement for better working conditions and for adequate compensation began in the mid-sixties when a small group of doctors, led by Donald Rassmussen, I.E. Buff, and Hawley A. Wells, began to explain to coalminers what they really knew already: that the coal dust was killing them slowly.[42] Almost from the start the black lung battle became connected to struggles for rank-and-file power and internal democracy within the union. "When we found out what was actually going on," said Bill Worthington, a leader in the black lung movement, "we began to get pretty angry. Our people were dying. They weren't getting benefits. Coal companies were making millions of dollars off of us, and then when we got too sick to work, they said we had 'miner's asthma' for which there's no compensation. You were just out. You went to the poor house or started begging." Then he said, "We began to get together . . . to do some serious thinking about what we could do."[43]

In 1969 while the West Virginia state legislature was in session, hundreds of miners marched on the state capitol, demanding a compensation law for black lung. At the same time, they closed almost all the mines in the state in a massive wildcat to support their legislative demands. Together, the strike and march forced the passage of the desired law. Meanwhile, Consolidation Coal Company's No. 9 mine near

Farmington, West Virginia, blew up and killed seventy-eight miners. The UMW leadership took advantage of public sympathy to press for a federal mine safety law. But the union leadership purposely ignored the issue of black lung compensation and occupational health, which it believed would jeopardize the passage of a safety law. So it was left to a remarkable coalition of rank-and-file miners and widows, progressive physicians, and liberal members of Congress, in open opposition to the UMW leadership, to push for black lung compensation and controls on coal dust.[44] This resulted in passage of the federal Coal Mine Health and Safety Act of 1969, a comprehensive bill dealing with black lung compensation, exposure to coal dust, and safety. It specified that coal miners, or their widows, could apply for permanent black lung benefits regardless of when they had quit the mines. Benefits were paid from federal tax revenues, and at first 60 percent of the applicants won the benefits. The law also created a federal inspection system to enforce safety and coal dust standards.

Passage of the Coal Mine Health and Safety Act gave new force to the Black Lung Associations (BLAs). Retired miners and miners' widows who had successfully secured benefits began to lead others through the Social Security maze, and soon constituted themselves into effective county-level organizations. These groups often got together on a state-wide or national basis for specific political battles, such as the Black Lung Benefits Act of 1972, and there was a large overlap between the BLA people and the Miners for Democracy, who were trying to take over the union after the murders of Jock Yablonski and his wife and daughter in 1969.[45]

The Black Lung Benefits Act of 1972 was the BLA's legislative baby. The idea was to relax the medical eligibility requirements for black lung benefits. Under the 1972 act a miner who had worked for fifteen years underground was presumed to have compensable black lung if he was totally disabled by lung disease, regardless of what x-ray findings might indicate. Even with its passage, however, the Social Security Administration managed to reduce the rate of claims success from 60 percent to under 10 percent. By the mid-seventies most of the blatant cases of black lung had already been compensated and the government and coal companies were increasingly reluctant to grant benefits.

Arnold Miller, the winning reform candidate for president in 1972, was himself a black lung victim, who had built much of his political reputation around his work with the Black Lung Associations. When he came to power he promised to make safety and health his "utmost concern." For the first time a reform movement took over a major union with health and safety as a prime issue.

Since Miller's accession to the presidency the miners have twice shut down most of the nation's underground mines over safety issues and over attempts by the authorities— management, union, and the courts—to curb the right to strike.

The right to strike over safety and other grievances was one of the principal issues separating the miners from the coal companies in the four-month strike in 1977-1978, the longest national coal strike in history. When the miners finally went back, proposed heavy sanctions against wildcatters were deleted from the proposed contract.[46]

Despite restrictions and limitations, however, the revolt over black lung in the last decade has channeled a tremendous amount of money to miners and their descendents. Gail Falk, a former UMW attorney, reports that over half a million families have received some financial benefit from the black lung program.[47] The annual payments of over a billion dollars a year are 20 times the total paid out for occupational disease for all other workers.[48] In other words, 95 percent of occupational disease benefits in the United States are paid to coal-miners and their families. For the first time in recent years the private sector lost control of workers' compensation, and company physicians have lost the right to determine what is a compensable disease.[49] Federal expenditures for research on black lung disease jumped from $100,000 in 1963 to $7 million nine years later.[50]

Conditions in the mines are probably safer and less dusty as a result of the new safety and health legislation. Bureau of Mines statistics claim that coal mine accident deaths fell from

260 to 132 between 1970 and 1974.[51] The new regulations also created new jobs, at least temporarily, as the companies hired people to comply with the law.[52] So long as demand for coal and labor is high and rising, miners can successfully demand better pay and conditions, and everyone concerned can look forward to more open class warfare in the coalfields.

International Association of Machinists (IAM)

In 1969 the national office of the IAM sent a directive to all local lodges to set up safety committees. The policy outlined in the *IAM Guide to Safety and Health Committees* is "to have *union-controlled* Safety and Health Committees in all plants and shops under IAM contract. These are to be *union committees*, solely under the control of the union and completely independent of any joint union-management committees that might also be in existence."[53] The chair of the local safety committee is registered with the national office. To back up these committees, each of the seven general vice-presidents around the country has a part-time safety representative. At headquarters in Washington, D.C., the IAM maintains a safety and health department. Dr. Thomas F. Mancuso, one of the two or three most respected occupational health physicians in the country, works half time for the program. He writes an excellent monthly question-and-answer column for *The Machinist* which goes to all members of the union. The best of these have been collected in a paperback called *Help for the Working Wounded.*[54]

Neither the regional nor the safety and health representatives provide training for local committee members on a systematic basis, although they do help set up regional training sessions. National Safety Director Angelo Cefalo has stated that he is now trying to organize a training program through OSHA if he can get the government to pay for lost wages during the course.[55] The Machinists' union, with nearly a million members in the highly hazardous metal-working trades, has a smaller national health and safety staff than the Oil, Chemical, and Atomic Workers, with one-fifth as many members. Given the wide range of problems on the local level, the minuscule staff, and the formal decentralization of authority in the Machinists, the action is mostly on the local level.

Local Lodge 284, which represents 4,000 workers at Caterpillar Tractor and other shops in San Leandro and Hayward, California, established a twelve-member health and safety committee in 1976. The members of the committee have already carried out a preliminary survey of the problems in their shops, and are acquiring further health and safety skills at a course with the Labor Occupational Health Program in Berkeley.[56]

At San Francisco International Airport, workers at the United Airlines maintenance base and some of the cargo handlers are organized into Local Lodge 1781. Some 8,000 workers do major maintenance work for United and over twenty other airlines, and for the most part they are highly skilled and paid, averaging over $16,000 a year in wages. The facility contains hundreds of different operations and job hazards, because the workers often disassemble and overhaul entire aircraft. However, the company concern for aircraft flight safety guarantees some protection from accidents caused by speedup. In order to ensure quality work, mechanics are individually responsible for the tasks they perform. If a boss tries to hurry the work on a particular piece, the person working on it can refuse to sign off the task and demand that the supervisor take responsibility. In such situations the supervisors almost invariably refuse.

In 1972 a new union president, John E. Stilwell, and a new slate of officers were elected. They carried out a complete review of the contract between United Airlines and Machinists District 141, of which Local 1781 is a part. Article 20 of the contract, which had never been interpreted or applied to the advantage of the union, reads as follows:

No employee will be required to work under unsafe or unsanitary conditions, and in order to eliminate as far as possible acci-

> dents and illnesses, a joint safety committee composed of an equal number of representatives designated by the union and by the company will be established at each location on the [United Airlines] system, where employees covered by this agreement are employed.
>
> It shall be the duty of the safety committee to see that all applicable state and municipal safety and sanitary regulations are complied with and to make recommendations for the maintenance of appropriate safety and sanitary standards. The committee shall receive and investigate complaints regarding unsafe and unsanitary working conditions, and make recommendations concerning such complaints. Union safety committee members shall be allowed a reasonable amount of time during working hours without loss of pay for these purposes.

Safety committee head Bob Fowler interpreted this to mean there would be a separate union committee and a separate company committee, and that a reasonable amount of time to investigate hazards meant full time. The only question to be resolved was the number of people to include on the union committee. So at 7:30 one morning Fowler strode onto the shop floor and began noting down violations. The director of industrial relations ordered him back to the carpentry shop and threatened to reprimand him for insubordination. Fowler stood his ground, so United finally requested to discuss the matter with the union. After two twelve-hour bargaining sessions the company agreed to eight full-time union safety committee members, at no loss in wages. Individual workers and committee members got the right to shut down any job they considered unsafe.

Over a thousand people who work at the airline terminal in food handling, cabin cleaning, cargo handling, and routine maintenance are members of Local 1781. For these workers the accident rate was extremely high. In 1974, of 150 people in air freight handling, there were 43 lost-time injuries: mostly sprains, back injuries, and hernias, which led to over 800 lost work days. Soon after the agreement to create eight full-time safety people from Local Lodge 1781, three of these were chosen from the terminal workforce, by verbal agreement with the terminal manager. In the first year and a half of the program there, the number of lost-time injuries was reduced by nearly 50 percent.

The union safety committee impressed management with its effectiveness in reducing accident costs, but its power and following among the workers has caused United Airlines to try to eliminate it. Instead of five full-time safety committee members at the maintenance base, there are now three. The committee force at the terminal has been cut back from three full-time people to two people at only four hours a week each, a reduction from 120 hours per week to 8. No one I talked to could explain what kind of deal the union president had made to account for such a severe cutback. According to one safety committee member I talked to in April 1977:

United Airlines would be happy if we just dried up and blew away. They were willing to tolerate three full-time safety people at the terminal, so long as we didn't challenge them on anything big. We ran into our first big fight from United when we finally filed a CalOSHA complaint about one of their maintenance operations called a "boroscope test" on the no. 2 engine of a DC10. In order to perform this test the mechanics had to cross a platform 20 feet high where they were in danger of falling. If they followed United's regulations in doing the test they could have been crushed by the engine cowling if the hydraulic system failed. Changing the procedure will cost in excess of $200,000— they would need new platforms, engineering, and would have to write new specifications for the job.

This person explained that the safety committee's experience with OSHA and CalOSHA (a California agency which has replaced OSHA) has been unsatisfactory. Most of the problems require constant vigilance and immediate solutions, particularly at the terminal, yet United Airlines routinely appeals all citations, which can take from fifteen months to two years to resolve.[57]

Local Lodge 1781 has also shared its experience in health and safety with other groups. It founded the Airport Coalition for Safety and Health at San Francisco International

Airport. The new organization took off "like a ball of fire" but it lacked a strong base. When the initial goal of a twenty-four-hour medical service for the airport was secured from the San Francisco municipal government, most of the participants dropped out. Management representatives, who had joined the coalition mostly to keep an eye on the active unionists, soon lost interest. One participant from Local 1781 remarked that in the future they would have to exclude management if they wanted to get anything done.

The experience of Local 1781 in health and safety has reached many people within the Machinists' union and the labor movement as a whole. Its biweekly newspaper *Tradewinds* has spread its experience throughout District 141, which represents dozens of United Airlines shops under the same contracts. As the largest local in the District, 1781 has been able to afford to send health and safety committee members to training sessions around the country. After its early interest in safety, 1781 won two National Safety Council awards as the best local union health and safety program in the country. The founder of the program, Bob Fowler, was hired by the Labor Occupational Health Program in Berkeley and wrote up his experience in the exhaustive *A Guide for Local Union Health and Safety Committees*, which is still available from LOHP.[58]

Officially, however, the Machinists' union, at both the national and regional levels, has ignored the accomplishments of Local 1781 in health and safety. A maverick local, 1781 has been in conflict with higher levels of the union on other issues. Though a staffer from *The Machinist* once did an extensive interview with Fowler about the program, it was never printed, and the district organization has never responded to 1781's offers to train other locals in health and safety, though they are covered by the same contract. The result of 1781's activities, nevertheless, has been to help create "a lot of upward pressure all over the United Airlines system and in Washington" for more health and safety services. The smaller locals want more training sessions and advice from the district, and 1781 is eager for more lobbying help in Washington to protect the OSHA law and ensure vigorous action on the Toxic Substances Control Act of 1976.

United Steelworkers of America (USWA)

I'm a dying breed. A laborer. Strictly muscle work . . . pick it up, put it down, pick it up, put it down. We handle between forty and fifty thousand pounds of steel a day (laughs). I know this is hard to believe—from four hundred pounds to three- and four-pound pieces. . . . A mule, an old mule, that's the way I feel. Oh yeah. See (shows black and blue marks on arms and legs, burns). You know what I heard from more than one guy at work? "If my kid wants to work in a factory, I am going to kick the hell out of him! I want my kid to be an effete snob." Yeah. Mm-hmm (laughs). I want him to be able to quote Walt Whitman, to be proud of it.[59]

Ken Bellet, a chief steward for the Steelworkers' union in the huge Bethlehem Steel plant in Buffalo, New York, tells horror stories about life in the mills which rival what muckrakers described seventy years ago:

A couple of weeks after I started working . . . a foreman was killed. A lift fell on his head and squashed him like a bug. One guy had a heart attack near the furnaces and died waiting for the car to take him to the dispensary. There was a man killed about a year ago—his sleeve caught in a round straightener and he got whipped to death. There's an Arab we call Snake. One day he had his arm on a pile of steel bars. The crane set a load on his arm and his arm ended up like a waffle.

Mechanics carry a blue flag, a safety flag. They put it on the switch while they're working on a machine, as a warning not to throw the switch. One time I was working on a straightener repairing a grease line. I had my flag on the main switch. I was reaching in between two rolls trying to get the grease line out. The foreman came over hollering to get that blue flag off the switch. It was holding up production. I told him if I didn't replace it the line roller would freeze and maybe a bar would fly off and hurt somebody. His comment to that was, "I don't give a good fuck," and he pulled the flag off the switch and turned the machine up with my arm between the rolls.[60]

The United Steelworkers of America (USWA) lobbied more than any other union for the OSHA law. In the autumn of 1970 their Washington representative, Jack Sheehan, worked nearly full time on the bill, and dozens of rank-and-file steelworkers came to testify and pressure legislators. The Steelworkers' leadership opposed Nixon administration attempts to substitute a weaker bill, and successfully countered efforts by Democrats and Naderites to compromise away the authority of the Secretary of Labor to set health and safety standards.[61] Since its passage, the Steelworkers have used the OSHA law as the main tool for securing better conditions in the steel industry. Legislative Director Sheehan has argued forcefully that "Many employers—particularly smaller ones—simply cannot make the necessary investments in health and safety and survive unless all are compelled to do so." It follows, according to Sheehan's argument, that collective bargaining pressure is basically unsuited to securing better conditions in the workplace. Most workers are not covered by collective bargaining agreements, and many of those who are lack the power to extract significant concessions. Worst of all:

> Through collective bargaining, safety might shift from being part of the employer's cost of doing business to being the price that workers would pay for having a job. In its more gruesome terms, it . . . has been . . . translated into . . . a trade-off between jobs and safety, and the burden to make the decision is placed upon the worker.
> Under OSHA, however, safety becomes a *social cost* of doing business rather than an *employment cost* which could, in collective bargaining, be expended for a variety of benefits. . . .[62]

Since OSHA's passage the Steelworkers' national leadership has strongly opposed crippling amendments in Congress, and regarded devolution of enforcement authority to the states as the greatest threat to the law's integrity.

The Steelworkers' effort in Washington was not immediately matched by the safety staff in Pittsburgh. At the founding conference I attended of the now-defunct Pittsburgh Area Committee for Occupational Safety and Health (PACOSH) in December 1972 it was clear from the questions that the Steelworker participants badly needed more information about health hazards. But the union's close coordination with the needs of industry, symbolized by the no-strike Experimental Negotiating Agreement (ENA), and its tradition of autocratic leadership have discouraged rank-and-file initiative.[63] The safety department has opposed the formation of independent groups such as PACOSH, which bring together people from different unions and technical specialties.

The national union's first major occupational health battle has been the fight against coke oven pollution. Coke is produced by baking high quality coal for sixteen to twenty hours at temperatures often exceeding 2,000° F. This process drives off most impurities and leaves a residue of almost pure carbon called coke, which is suitable for use in blast furnaces for steelmaking. The greatest exposure to harmful substances occurs when the coal is charged into the oven, and when the red-hot coke is pushed out. The (mostly) black workers who have been forced to work on top of the coke ovens contract lung cancer seven times as often as the general male population, and they also have high rates of other lung diseases.[64]

Since the mid-sixties there has been a strong local movement to clean up the coke ovens. Daniel Hannan, a former president of USWA Local 1557, in Clairton, Pennsylvania, helped found GASP (Group Against Smog Pollution), to fight against emissions from the Clairton coke ovens, the largest in the nation. In 1974 U.S. Steel, the owner of the Clairton works, was pressured into signing a special agreement to hire 172 new workers to improve maintenance procedures and cut down exposures in the environment and at the workplace. According to Charles Stokes, grievance committee head for the coke workers, the improvement is "like the difference between night and day" over the past eight years, though there is still a great deal to do. Daniel Hannan was hired onto the Steelworkers' national safety staff.[65]

After the passage of OSHA the national staff of the Steelworkers mounted a major campaign for a strong coke oven standard. To aid the effort the safety department hired Dr. J. William Lloyd, author of the definitive occupational health studies of steelworkers, and Claudia Miller, an industrial hygienist, to provide technical support. Rather than relying solely on the setting of a threshold limit value for airborne levels of pollutants, the standard also requires specific work practices. After each "push" of red-hot coke, for example, the oven doors must be cleaned to prevent the build-up of tar. The prior practice of cleaning the doors after every ten pushes made it impossible to close them tightly, leaving the air near the ovens full of poisonous smoke.[66]

The Steelworkers' union, which according to hygienist Claudia Miller spent half a million dollars on the coke oven standard, has been criticized for pushing for new technologies to eliminate coke oven emissions. According to the Steelworkers' leadership, clean technologies for producing coke exist in the USSR and Japan. There, closed pipeline systems are used to charge the ovens with coal, eliminating the major source of pollution. Instead of quenching the hot coke by pouring water on it in the open, it is cooled slowly in closed chambers which trap much of the heat for power generation.[67] Although this technique might pay for itself by producing a better grade of coke, U.S. steel companies have ignored it. The problems with the new coke oven standard, according to Steven Wodka of the OCAW health and safety staff, are that the eight-hour exposure limit is still set at a level which will cause some cancers (though the level is lower than at most present operations); the companies are given until 1980 to implement the new standard; and most importantly, the standard ignores the issue of rate retention—equal pay at a less hazardous job for those whom coke oven work has already weakened.[68] Although the union promised to fight for full rate retention for all dangerous jobs in the 1977 contract talks, rate retention was secured only for those in the lower pay classification, and only if the company doctor agreed.[69]

Another area of union activity is lead poisoning, which, in its milder forms, causes listlessness and depression and in high concentrations can be fatal. Long-term exposure often makes men lose their desire and ability for sexual intercourse and can cause both women and men to become sterile, or to have babies with birth defects. Even low-level exposure causes sperm damage.[70] The USWA has taken the lead in arguing for much stronger lead standards than those proposed by OSHA, and has brought in dozens of rank-and-file members to testify for them at the OSHA hearings. The goal of the Steelworkers' proposed standard is to keep lead levels below 30 micrograms per 100 milligrams of whole blood, the upper range for an urban American who does not work near lead. This is also the level at which sperm abnormalities begin to appear, according to a Rumanian study.[71] The Steelworkers seek to mandate that workers suspected of exceeding this level must be given immediate blood lead level tests and be given the option of donning individually fitted forced-air respirators. Under this standard, women workers of childbearing age would have the option of transferring out of areas where their blood lead levels rose over 30 micrograms, with full rate retention. Men could transfer out when their blood lead level hit 50, also with rate retention. The Steelworkers also requested that the time-weighted permissible exposure to lead be set at 40 micrograms per cubic meter of air, rather than OSHA's proposed standard of 100 micrograms. The present standard is 200 micrograms.

Women's groups, such as the Coalition for the Medical Rights of Women, have testified strongly against any differences between legally permissible exposures for women and men, because companies use the differences as an excuse to get rid of women workers. At the Bunker Hill smelter in Kellogg, Idaho, few women still work at well-paying jobs with high lead exposure. According to the testimony of steelworker Eva Sullivan, the company feared liability suits from the parents of defective children.[72]

It took five and one-half years to pass the coke standard,

and the battle for implementation is just beginning. Daniel Hannan of the health and safety staff believes it will take ten years to make the steel industry comply with the new standard. The union has already held several coke oven conferences, attended by at least 100 people each, and has begun to use training manuals and slide shows from Ohio State's OSHA Research Project. There is talk of developing more educational programs and creating a Steelworkers' school which would include health and safety among its subjects, but so far the union relies heavily on official OSHA publications for training purposes.[73] The only union pamphlet written for a large audience, *USWA Safety and Health Program* (1976), is rather conservative, as it presupposes a joint management-union safety committee at the local level, rather than an independent union committee.

If the union is to deal adequately with the problems not only of coke oven workers but of the other 98 percent of its members, it will have to rely much more heavily on rank-and-file talent and initiative; there is no way the staff can meet the demand. Steelworkers' industrial hygienist Claudia Miller has complained:

> Bill Lloyd and I are spread very thin and we are very tired. We are a drop in the bucket compared with what is really needed, and they demand more and more of us. To give you an idea, our people at Dow Chemical in Midland, Michigan sent us a 40-page list of the chemicals used there, all trade names, and asked us to identify them and report back on the hazards. We are up against dozens of Dow professionals at that one plant, and that is just one local out of 5,500 in the whole union.[74]

A process of decentralization and local initiative has already begun in some locals. A foundry local lobbied with the Connecticut state legislature to ease compensation criteria for silicosis victims[75] and the giant Bethlehem Steel locals at Sparrows Point, Maryland, have initiated struggles on a wide variety of job health issues.[76] Relations are cordial between CACOSH and the safety department of the Steelworkers in

Pittsburgh. Despite problems of excess capacity and unemployment in steel, the issue of working conditions will continue to grow in the industry.

United Electrical Workers (UE)

The United Electrical, Radio, and Machine Workers was formerly the third largest industrial union in the nation, with 600,000 members. During the post-World War II red scare the government and big business saw the UE as a threat because it was a militant and democratic union and because it initially refused to follow the cold war line in foreign policy or expel Communists and other radicals from its ranks. In order to forestall attacks by other unions in collusion with employers, the FBI, and Senator Joe McCarthy, the union had to authorize its officers to sign non-Communist affidavits. Even so, the raiding by other unions continued. At one point the UE sank to a low of 90,000 members before it began to grow again.

Throughout its difficult days in the forties and fifties, the UE retained its fighting democratic traditions. The UE constitution requires that the pay of the leaders shall be equivalent to that of the most skilled workers in the shops, so that they live at the same material level as the rank-and-file. In 1972, for example, the three top officers of the union were allowed to make a maximum of $284.14 per week, international representatives $215.14 per week, and field organizers $197.60. Yet the leadership refuses to press for higher salaries. Instead, James J. Matles, general secretary, argues how well he is doing in the following terms to rank-and-file convention delegates:

> We officers, organizers, business agents, district presidents, have it all over you as far as the job is concerned. When you walk through that plant gate every morning, most of you hate to do it. If you did not have to earn a week's pay, few of you would ever go near that gate. During the years you have kept me on my job

I've been able to stay away from that gate 9,000 times. Instead of going to work every morning for a boss, and hating it, I've been getting up every morning and going in to work at a job I like to do. That's where we officers have it all over you.[77]

In keeping with its traditions, small size, and the diversity of its members' problems, the UE has stressed the importance of the education of its members on occupational health problems so that they can solve problems by themselves. They invited ecologist Barry Commoner to address a national convention on the links between a healthy environment and a healthy workplace, and published his speech in a pamphlet.[78] On technical questions the union often retains Dr. David Kotelchuk, a biophysicist and writer for the Health Policy Advisory Group in New York. The *UE News* reports the activities of its locals on occupational health and the activities of grassroots groups such as CACOSH. In addition, union members work with these groups. Frank Rosen, international representative from Chicago, was one of the first officers of CACOSH, and Frank Collonello, chief steward of UE local 610 in Pittsburgh, was the first president in PACOSH in Pittsburgh. UE Local 108 at the Gould Battery plant in Trenton, New Jersey, went on strike to implement a lead control program developed with the Health Research Group in Washington, D.C., and has worked closely with PhilaPOSH in Philadelphia.[79]

Progress on health and safety is restricted in many workplaces because of the structure of the industry. Although the UE has managed to retain a small foothold in the giants of the electrical equipment industry, such as General Electric and Westinghouse, most of its members work for small shops where employers face keen competition from nonunion shops and large competitors. If the union pushes the small shops too hard for better conditions, it risks driving these employers out of business. To fight back, the UE devotes a great deal of effort to organizing new shops at large enterprises which have fled to the South and Southwest to avoid unionization.

Amalgamated Clothing and Textile Workers Union (ACTWU)

The ACTWU is another union which has begun to use the occupational health issue to organize workers, devoting a good deal of attention to the problem in its organizing literature. It has given both financial and moral support to the Brown Lung Associations that are trying to secure compensation for victims of cotton dust disease in the Carolinas.[80]

Over the years the union's original base in the Northeast has eroded as factory owners have escaped to the low-wage, anti-union South. The textile industry in the United States is still very competitive. The biggest operator, Burlington Mills, controls only 7 percent of the total production, and many firms are still family-owned and operated with equipment that is forty or fifty years old. Without investment in new technology, it is possible to earn a profit only by keeping wages down, eliminating competition from foreign goods through higher tariffs, and preventing unionization.

The principal target of the ACTWU southern organizing drive has been J.P. Stevens, the second largest textile producer in the nation. The company employs 44,000 workers in eighty-five plants, mostly in the Carolinas. Only one of these plants is unionized, and even there J.P. Stevens has refused to sign a contract. The company routinely tries to fire workers sympathetic to the union, and often it succeeds. On one case, which took two and one-half years to win in federal appeals court, J.P. Stevens had to reemploy sixty-nine fired workers and pay $684,756 in back wages. A standard tactic is to threaten to shut down a factory if the union is voted in, and the threat has been carried out often enough in the South to make it believable. Following the United Farmworkers' example, the ACTWU is committing funds to organize a consumer boycott against J.P. Stevens to compensate for its weaknesses in the workplace.[81]

In late 1975 the union filed a suit against OSHA to force it to develop a new standard for cotton dust. The union has argued that exposure to high levels of cotton dust will be resolved in the long run only by reequipping much of the industry in the next few decades with productive, dust-free machinery. This machinery has already been introduced in the United States, and is widely used in both Eastern and Western Europe. It should enable the textile industry to survive the competition from cheap-labor countries and pay higher wages to the remaining workers. Unable to fight for every job in unorganized shops, the union reasons that it has nothing to lose if half of the jobs in the industry disappear in the next few decades, so long as most of the people in the new jobs are unionized at relatively high wages. At present only 10 percent of the 700,000-worker textile industry is unionized, because employers in this competitive industry still oppose the right of the union to exist.

In cotton dust standards testimony the union has called for strong OSHA-sponsored programs of medical surveillance, and wage retention for workers already too sick to work in dusty areas. Although chances are slim that OSHA can be convinced to influence long-range corporate investment decisions and to socialize medical decision-making in occupational health, the ACTWU proposals represent a first step in that direction. As one staffer put it: "These companies have to innovate or shut down. The unions and OSHA just might be the only forces capable of dragging this industry into the twentieth century."[82]

United Farmworkers of America (UFW)

The combined use of giant tractors, mechanical harvesters, and petroleum-based insecticides and plant poisons on large American farms has made over 5 million farmworkers superfluous since 1945, and subjected those remaining (and perhaps their descendants) to epidemics of new kinds of accidents and disease. There is hardly any government or private support for poison-free methods of pest control, despite the fact that all poisons eventually lose their effectiveness as insects become resistant.[83] Crops as well as workers must fit the machine and the changes are often as threatening to the palate as to the workers' health. For example, agricultural geneticists have designed a tomato for mechanical picking which can be dropped from six feet onto a tile floor without splitting. Grown in biologically dead soil and mulched with plastic, the tomato plants consume 1.5 tons per acre of artificial fertilizer and must be sprayed two or three times a week with fungicides, herbicides, and insecticides. Rather than ripening on the vine, these tomatoes are "degreened" in warehouses by exposure to ethylene gas. And for what? "Insipid," "blah," "tough," "like eating cardboard," and "plastic junk" are some of the most common descriptions of the pink tomatolike objects in the supermarkets today.[84] But chemical poisons are not the only hazards to farm workers.

According to National Safety Council statistics, the death rate for farmworkers is as high as for miners and construction workers, or four times the average for all occupations.[85] However, farmworkers are among the lowest paid workers in the United States. The average life expectancy of farmworkers is less than fifty years, compared to a national average of seventy. For this reason, often whole families work in the fields. The children are up at 4:30 in the morning to help their parents pick, and they often return after school to work until 7:30 at night. These are the words of farmworker Robert Acuna:

The hardest work would be thinning and hoeing with a short-handled hoe. The fields would be about a half-a-mile long. You would be bending and stooping all day. Sometimes you would have hard ground and by the time you got home, your hands would be full of calluses. And you'd have a backache. Sometimes I wouldn't have dinner or anything. I'd just go home and fall asleep and wake up just in time to go out to the fields again.

I began to see how everything was so wrong. When growers can have an intricate watering system to irrigate their crops but they can't have running water inside the houses of workers. . . . They can have land subsidies for the growers but they can't have adequate unemployment compensation for the workers. . . . In fact, they treat their implements better and their domestic animals better. They have heat and insulated barns for the animals but the workers live in beat-up shacks with no heat at all.

Illness in the fields is 120 percent higher than the average rate for industry. It's mostly back trouble, rheumatism and arthritis, because of the damp weather and the cold. Stoop labor is very hard on a person. Tuberculosis is high. And now because of pesticides, we have many respiratory diseases.

The University of California at Davis has government experiments with pesticides and chemicals. To get a bigger crop each year. In 1964 or '65 an airplane was spraying these chemicals on the fields. Spraying rigs they're called. Flying low, the wheels got tangled on the fence wire. The pilot got up, dusted himself off, and got a drink of water. He died of convulsions. The ambulance attendants got sick because of the pesticides on his person. A little girl was playing around a sprayer. She stuck her tongue on it. She died instantly.[86]

Most farmworkers are members of racial and ethnic minorities who can be forced to work for extremely low wages because they have nowhere else to go. The largest corporate farms in California usually hire Chicanos, foreign-born Mexicans (often without documents), Filipinos, and Arabs. (Nagi Daifullah, a twenty-four-year-old Yemeni Arab, was the first California farmworker killed by police in the long struggle for union recognition.[87] The growers have fought hard to keep their workers unorganized and rootless, following the crops with no place to call home. As a result, they are not protected by most provisions of the welfare state. Less than half of the states include farmworkers under worker's compensation legislation, so they are defenseless if injured or poisoned, and no state covers them for unemployment. In addition, minimum wage laws are less stringent for farmworkers than for other workers.[88]

The first serious attempts to organize agricultural workers under U.S. jurisdiction occurred in 1909 among Japanese sugar plantation workers in Hawaii. Their month-long strike was broken, partly by a failure of Japanese and Filipino workers to cooperate with each other. There were repeated strikes of sugar and pineapple workers for the next forty years. In 1920 a 165-day strike on Oahu did not secure union recognition, but wages rose 50 percent as a result. A 1924 strike of mostly Filipino workers was smashed only when the National Guard intervened. Sixteen workers and four policemen were killed, and from then on the growers made it a point to recruit illiterate workers from the Philippines. Hawaii's agricultural workers were organized only when the International Longshoremen's and Warehousemen's Union (ILWU) opened full-scale organizing campaigns; after a series of bloody battles, the ILWU won its first contract in 1941. With the passage of Hawaii's "little Wagner Act" in 1945 agricultural workers acquired the same rights to organize unions as other workers had under the 1935 Wagner Act. The ILWU quickly grew from under 1,000 to over 30,000 members, mostly on the plantations.[89]

Struggles to organize agricultural workers have been slower on the mainland. Sharecropper unions organized in the South in the 1930s were defeated by farmowners and the local police,[90] and farmworkers were excluded from coverage by the Wagner Act. They have begun to successfully organize only since the early 1960s, under the leadership of Cesar Chavez' United Farmworkers in California. The UFW's vision goes beyond the hourly wage; it would like to control hiring through a hiring hall and seniority, and enable farmworkers and their families to settle down and take part in community life in a dignified manner. One farmworker, now a union organizer, talked about what farmworkers want:

Working in the fields is not in itself a degrading job. It's hard, but if you're given regular hours, better pay, decent housing, unemployment and medical compensation, pension plans—we have a very relaxed way of living. But the growers don't recognize us as persons. That's the worst thing, the way they treat you. Like we have no brains. Now we see they have no brains. They have only a wallet in their head. The more you squeeze it, the more they cry out.

If we had proper compensation we wouldn't have to be working seventeen hours a day and following the crops. We could stay in one area and it would give us roots. Being a migrant, it tears the family apart. You get in debt. You leave the area penniless. The children are the ones hurt the most. They go to one school three months in one place and then on to another. No sooner do they make friends, they are uprooted again. So when they grow up they are looking for this childhood they have lost.[91]

When the owners of California's giant farms first discovered that they couldn't drive the UFW from the fields by themselves, they tried to bring in the Teamsters, with their violent goons, to organize in their stead. After years of organizing, consumer boycotts, and strikes across the nation, however, the Farmworkers finally forced the government, growers, and Teamsters to accept them as the legitimate agricultural workers' union in California.[92]

Throughout these years of struggle, occupational health and safety has been one of the principle organizing issues.[93] The union convinced the state of California to outlaw the use of the shorthandled hoe that had caused so much back trouble. For the bosses, the shorthandled hoe functioned as an instrument of social control. They had only to see who was bent over to tell who was working. The UFW also has negotiated some precedent-setting contracts to regulate the use of pesticides. One contract totally bans the use of the herbicides 2,4-D and 2,4-5T and the insecticides DDT, DDD, Aldrin, Dieldrin, and Endrin. Workers are not supposed to reenter any field treated with Parathion in less than twenty-one days unless essential work has to be done, and then protective clothing must be worn. The head of the union health and safety committee must be consulted in the formulation of policies related to health and safety and on the use of economic poisons. In addition, the company is required to keep and make available to the committee and to any worker the records on the use of hazardous substances. The contract also requires adequate rest periods, drinking water, and toilet facilities. But as a practical matter the UFW has been too preoccupied with its survival as an organization to maintain an ongoing program of education and action around health and safety. Nationally, less than 1 percent of farmworkers are organized—with hardly any of them outside of California and Hawaii. Even in California union members constitute only a small minority of all farmworkers.

The passage of OSHA has scarcely helped the cause of health and safety, as OSHA's Agricultural Advisory Committee is controlled by the growers, and OSHA's general industry standards do not apply to agriculture, despite the increasing use of machinery. Except for regulations requiring rollbars for tractors in case they tip over, and some weak standards which relate to migrant labor camps, the dangerous new machinery in agriculture is almost completely unregulated. Moreover, OSHA has neatly avoided the pesticide question: under grower pressure the organization rescinded a temporary emergency standard to limit worker exposure to pesticides, and turned over the whole problem to the Environmental Protection Agency (EPA). EPA's standards are not enforceable by law, and are based on pesticide manufacturers' "suggested" exposures, rather than on independent EPA research. To top it off, Congress has exempted farms which employ under ten workers from federal OSHA inspections for a year.[94]

In California a separate state law (CalOSHA) has preempted the enforcement of OSHA legislation from the federal government; this provides that agricultural workers be covered by the same health and safety standards as the rest of the workforce. Yet academic studies and state oversight hearings have shown that California farmworkers receive very little protection from the state inspection apparatus. Most of the licensed pest control advisors are pesticide salesmen who overload the land with poison. Enforcement of CalOSHA on the farms is carried out by county agricultural agents, whose primary ties are to the growers. A specialist from the State Department of Health found that only 1 or 2 percent of all pesticide-related illness is reported to the workers' compensation system.[95] In one case reported in state senate hearings, several farmworkers suffered severe body rashes from what they thought was exposure to pesticides. When they reported the event to the county agricultural agent, most of the sick were arrested the next day by the dreaded "Migra" (Immigration Service) and apparently deported, according to the *CalOSHA Reporter* of November 29, 1976. As one worker put it: "That's why farm workers are afraid. If they are illegal, they are subject to deportation. Even if they are not, little is done for them."

There is no inherent conflict between the needs for healthy working conditions and for jobs in agriculture. But there will have to be a large research effort and a painful transformation in farming practices to develop safe methods to control pests and crop diseases. At first the petroleum-based poisons seemed like a dream come true to farmers who had watched insects ruin their crops. But pesticides are losing their allure. As harmful insects develop a resistance to specific poisons the dosage and frequency of application must be increased, at tremendous cost, and farmers must turn to new and deadlier poisons. Neither pesticide manufacturers nor their allies in the Department of Agriculture have seriously considered non-poisonous methods of pest and disease control.

In the long run poisons will have to be almost totally elim-inated from the fields. They will be replaced by a variety of techniques called "integrated pest management," such as the interspersing and rotation of many different crops, the en-couragement of the natural predators of harmful insects, and the development of resistant crop strains. All this is old hat to organic farmers, but is only now being talked about by the Department of Agriculture. The indiscriminate sowing of poison by aerial cropdusters, now a growing business, would also have to be strictly regulated, if not discontinued alto-gether.

The new methods will certainly be safer for the workers and increase the number of jobs. With smaller fields, devoted to a mixture of crops, it will be harder to justify the purchase of gigantic harvesters which are suitable for only one type of crop, and are extremely expensive to buy, fuel, and maintain. Thousands of new experts who specialize in the intimate relationship of harmful insects and plant diseases with birds, other insects, the crops, and the weather of a particular locality will eventually replace the chemical company sales representatives. And the farmers, the workers, and the public will have everything to gain. The major losers will be the petrochemical industry's owners.[96]

Building and Construction Trades

Building and construction work is both hard and dangerous, requiring work outside in all kinds of weather. Construction workers usually make more per hour than factory
workers, but they often work on the dead run. The resulting
pressure to meet deadlines is a constant threat to health and
safety. Often workers identify with management's productivity goals so completely that they are their own worst enemies. When times are hard the boss threatens to replace
slackers; when times are easy they rush to complete the work
so they can start new projects. In spite of the pressure, or
perhaps because of it, construction workers are proud and
hardy. For some the joy of building is tremendous.

> Once a job has risen thirty feet or more above the ground, there
> are breezes to help dissipate the summer heat, and there are fewer
> gnats and mosquitoes. There is none of the mud that gets be-
> tween the treads of a man's boots to compromise his footing. The
> other trades are left below (except, on a core job, for those core-
> finishing lathers, carpenters, and concrete workers who are
> unavoidably omnipresent). The air is cleaner. A fellow begins to
> feel less like an ordinary laborer doing the same things that every
> other ordinary laborer is doing, and more like an ironworker. But
> best of all, things finally begin to move a little faster, and one's
> sense of accomplishment is thereby expanded.[97]

Like underground mining, construction is inherently more
dangerous than most kinds of factory work. The work site
changes constantly, and often workers hire out to dozens of
different bosses in a single year. The danger sharpens the
feelings of self-esteem for some; others take to drink or
become insane, but the fear is there, even if people do not
talk about it.

> There's no job in construction which you could call an easy
> job. I mean, if you are out there eating dust and dirt for eight, ten
> hours a day, even if you're not doing anything, it's work. Just
> *being* there is. . . .

The difficulty is not in running a crane. Anyone can run it. But making it do what it is supposed to do, that's the big thing. It only comes with experience. Some people learn it quicker and there's some people can never learn it. (Laughs.) What we do you can never learn out of a book.

There's a bit more skill to building work. There is a boom crane. It goes anywhere from 80 feet to 240 feet. You're setting iron. Maybe you're picking fifty, sixty tons and maybe you have ironworkers up there 100, 110 feet. They're working on beams, anywhere from maybe a foot wide to maybe five or six inches. At the same time, they're putting bolts in holes. If they want a half-inch, you have to be able to give them a half-inch. I mean, not an inch, not two inches. Those holes must line up exactly or they won't make their iron. And when you swing, you have to swing real smooth. You can't have your iron swinging back and forth, oscillating. If you do this, they'll refuse to work with you, because their life is at stake.

There's a certain amount of pride—I don't care how little you did. You drive down the road and you say "I worked on this road." If there's a bridge, you said "I worked on this bridge."

That building we put up, a medical building. Well, the granite was imported from Canada. It was really expensive. Well, I set all this granite around there. So you do this and you did it good. Where somebody walks by this building you can say, "Well, I did that!"[98]

One of the prime tasks of the Ironworkers' union in New York City has been to place some limits on the employers' ability to decide on suitable working conditions.

Before the unions developed strength the employers called all the shots, and a man did what he was told or was replaced. If the company told a gang to erect iron in the rain, it was erected in the rain. Erection is never a safe job, and in the rain it becomes a great deal more hazardous, but a man did it under whatever conditions prevailed, because if he didn't, someone else got his job. In the dark hungry days of the thirties, gangs of out-of-work ironworkers hung about around job sites, so that when a man fell, they would be instantly available to take his place. The pay of a man who fell was stopped at the time of his fall. Today things are better. When a man falls today, his widow and children are paid the full day, even if he fell at 8:15 in the morning.[99]

Not uncommonly, the union waives safety precautions. Bob Fowler, a carpenter who later became very active with the Machinists in health and safety, helped dig out four Mexican laborers who were buried alive in Orange County, California, in a sandy ditch because management refused to shore it up. The next year he saw another horrifying accident, which he related to me as follows:

In 1963 I was working in Santa Ana, and there was always this thing about speedup. Everybody was working piecework, and everybody was making a lot of money and all the rules were being broken.

It was early one morning and there was a light mist. It had rained the night before and it was damp. We were framing tract houses and there was a crew on the house behind us nailing up plywood sheeting on the roofs. It was right after the first break, and this guy climbed back up on the roof and grabbed his [circular hand] Skil saw. He had the guard wedged open 'cause it's faster, with about half the open blade exposed. When he pulled the trigger to activate the saw, he couldn't let go of it. He was getting shocked, and he started yelling and dancing around. Immediately someone heads for the power pole when that happens because you've got to shut the power off—somebody's got a hot saw. But before anybody could get there, he had ripped himself open from the shoulder down to the hip, right across the front of his chest, because when you're getting shocked you become part of the saw and can't let go. It killed him right on the spot; there was no way he could have survived.

We actually had to put him on a pallet and then they lowered him off the roof with a fork lift.

That was the first time I really related anything to safety violations. He had violated a safety rule. He wedged the Skil saw open and you're not supposed to do that, but everybody did it, mainly because no one enforced the rule. Foremen would turn their heads; union stewards would turn their heads. Because there was so much work nobody seemed to care any more about safety violations, and I think if anyone was ever to check they'd find out that during those years of mass construction in the early sixties they'd find that the death rates and accident rates were two or three times what they should have been, based on years of unemployment. I said to myself, "If these people would just enforce their own rules, that probably wouldn't have happened.[100]

Union leaders in the construction trades are concerned that there have been *too many* inspections. Robert A. Georgine, president of the Building and Construction Trades Department (BCTD) of the AFL-CIO wrote: "In our opinion over the past five years, OSHA has conducted a disproportionate number of inspections on union jobs. This has reduced the competitiveness of the union contractor over the nonunion contractor. . . . OSHA should be called in as a last step rather than the first."[101] To take the pressure off union contractors, the Building and Construction Trades Department advocates more random inspections (rather than responses to complaints) so that nonunion contractors (whose workers are less likely to file OSHA complaints) will be forced to raise their safety costs. But there has been no advocacy of an increased inspection force. In order to cut down member complaints to OSHA, Georgine recommends that construction workers take their complaints to the union so it can deal directly with management. Where local union procedures to correct safety and health violations do not exist, Georgine recommends the following:

Steps to Follow When a Hazard is Observed:

1. If the hazard is within the authority of the draftsman, he should take steps to solve the problem himself. This may include talking to the craft foreman if appropriate.

If the situation is not corrected:

2. Bring the hazard to the attention of the [union's] craft steward. The steward should try to correct the problem by working with job supervisors.

If the situation is not corrected:

3. The craft steward should inform his Business Agent. The Business Agent should try to resolve the issue with the Project Management. Failing this, the B.A., depending on Local Union policy, should inform his Business Manager.

If the situation is not corrected:

4. The Business Manager, depending on the nature and imminency of the hazard, should bring the matter to the Local Building Trades Council. The Council should meet with the Project Management to resolve the situation.

If the situation is not corrected:

5. With the recommendations of the Building and Construction Trades Council and Local Unions, a formal OSHA complaint should be filed with OSHA.

If OSHA fails to correct the situation:

6. The Building and Construction Trades Department and International Unions should be asked to bring the matter before the Secretary of Labor through the appropriate area, regional or national office.[101]

These are hardly instantaneous remedies in the fast-moving construction business; if they were followed, local action would be almost impossible. At no times are workers advised to strike or refuse dangerous work. It is hard to imagine ironworkers going through this process in trying to rid themselves of an unsafe crane operator.

The philosophy of the Department of Safety and Health of the Building and Construction Trades Department is indistinguishable from that of management. Indeed, the BCTD generally lines up behind the construction industry in its lobbying over OSHA standards. The BCTD's manual, *Construction Safety and Health* (developed under a $753,000 OSHA grant),[102] treats its "first program objective" as the promotion of "construction safety and health as a cooperative effort between labor and management." It relies almost entirely on the construction industry and the National Safety Council for information. Asbestos is never mentioned as a problem, and personal protective equipment is treated as a solution to exposure to hazardous dusts and noise, rather than as a stopgap until better tools and processes become available. The best part of the BCTD's OSHA course is its training in hazard recognition and documentation.

The Department of Safety and Health of the BCTD was directed and staffed by one person in 1977, and it did not produce a newsletter or make available other educational materials.[103]

The Limits of Union Action

Although workers often express deep concern for health and safety on the job, they rarely act openly to change conditions. Many believe that the given situation cannot be changed, and others are concerned that enforcement of safer work practices will reduce productivity and pay, particularly if they work on a piecework basis. For many, resistance to unsafe work practices takes the form of sly sabotage, of letting machines break down without calling maintenance in time.

There are inherent difficulties in raising the occupational health issue with workers, particularly when the hazards are from low-level long-term exposure to hazardous substances. Usually people at work have been kept unaware of the serious dangers of industrial toxins, so where the unions fail to educate their members about these hazards, the workers never find out, unless they are extraordinarily persistent and can research the matter on their own initiative. Another problem is that workers at a particular job or factory may be exposed to dozens or hundreds of different hazards, each affecting only a few workers. Moreover, since most industrial chemicals are not tested or labeled in a meaningful way,[104] it is impossible for workers to realistically assess the extent of the problems they face. Even then, the educational problems are enormous. The relative success of coalminers in securing compensation for black lung disease has been due to the obvious and widespread nature of the hazard, once the effects of coal dust were well understood.

Behind much of the reluctance to agitate for better conditions is a fear of unemployment. In recent years unionized construction workers have been particularly concerned about losing their jobs to nonunion labor. In Wisconsin the unions have opposed such OSHA requirements as guarding of roof perimeters, despite numerous fatal falls, because this would slow down the work and unduly increase contractor costs. Since the chance of being killed or seriously injured by an accident on the job is usually less than one in a hundred in a given year, and since the long-term occupational disease risks are kept secret, workers tend to believe "it can't happen to me." Others laugh off the risks. In a standard joke among punch press operators a worker holds up two fingers and asks, "What's this?" The answer is: "A department 32 operator at A. O. Smith ordering four beers."[105]

Often, divisions emerge between workers in different jobs or areas of a plant, because they face different hazards. The younger workers who do the dirtiest jobs are much more concerned about safety than those who have graduated to softer work. One younger worker at A. O. Smith in Milwaukee attested that:

> Older workers resent the younger workers. They think safety consciousness is laziness. The older workers take the easiest jobs that their seniority entitles them to. For example, on the frame line, they take the repair jobs where you're not right over the smoke all the time and there's good cross-ventilation. You can stand back away from it, whereas in the fixture . . . the smoke comes right up in your mask, no two ways about it. . . . The younger man has to work at breakneck speeds to meet the incentive system. The older man just can't take that kind of work and he won't. He'll bump some younger man or go into another area—inspection, handyman on the line—then he doesn't care about slowing the line down. He wants his bread and wants the younger man to make it for him. Kids will do anything to get out of fixture, even fake a sprained wrist, fake a limp; but there is a six-year wait.

Another time the same company solved one safety problem by creating another:

> On our line we fought to engineer safety. At one time we ran into
> 17 people with back sprains . . . on one fixture. It was a funny lit-
> tle front end for a Chevy Station Wagon. Couldn't grab ahold of
> it anyway. People were hurting themselves.
>   At changeover time we got them to put in a nice easy slope.
> They said, "OK, we put in a safety device for you. Now we're
> going to cut your rate. So what they did was nothing because
> now the person's gotta go twice as fast to keep up with the line.
> The poor guy who is pulling this thing from a dead stop at the
> other end of the line has to pull twice as hard, twice as fast. This
> was the company's response to a safety grievance that we won.[106]

Adding to the problem is an inherent conflict between
rank-and-file members and the union hierarchy. Health and
safety problems are part of the daily grind for most workers.
But unless a catastrophe or wildcat strike occurs, the union
leadership pays little attention to working conditions. Labor
leaders have become administrators and enforcers of labor
contracts. They are part of a complex bureaucratic environment in which they must surround themselves with more and
more lawyers and accountants. As Stanley Aronowitz has
noted, their survival depends on their ability to maneuver
between a balky and demanding membership and the corporations and government.[107] National leaders often identify
strongly with the problems of corporation management;
increasingly, national union leaders are closer to their corporate counterparts in attitude and lifestyle than to the
membership.[108] In the clothing industry the International
Ladies' Garment Workers Union has even begun to teach the
employers how to run their small businesses, in order to save
them from foreign competition. As one labor journalist
expressed it:

> These guys . . . can enunciate these conditions, how shitty it is,
> but their commitment to change is a good deal less than total. . . .
> First of all there is the fact of estrangement. Even the negotiation
> committee does not work in the plant. . . . They don't know what
> a bitch it is hanging doors. . . . Labor leaders define their situa-
> tions as guys who have gotten away from this kind of shit.[109]

One result is an inherent tendency on the part of top union leaders to favor wage increases over nonmoney demands. Wage increases enable the organization to raise its dues without any corresponding increase in staff work. Similarly, increases in employer pension fund contributions give the union's leadership more financial power; depending on the degree of control union trustees have over the funds, leaders can raise their own salaries and put more of their allies and relatives on the union payroll. By contrast, the occupational health and safety issue confers no direct benefits on union officials. Instead, it increases the staff workload without generating additional income to deal with the demands on staff time. New and untrustworthy health technicians who want direct contact with the rank-and-file must be hired. Since safety is inherently a local issue, it shifts power to "hotheads" who are the ". . . natural enemies of the union official, [with a] disconcerting tendency to assume that every grievance must be settled here and now."[110]

Health and safety creates areas of potential conflict which can destroy a smooth relationship between management and the union. All the benefits of bringing about a healthy and safe workplace accrue to those outside existing union staffs: to the workers themselves and to the experts who furnish research and advice—for a price. For the United Mine Workers the occupational health and safety issue was a major factor in a successful takeover bid. It is no wonder that top union leadership has trod gingerly. The day-to-day work on health and safety questions at AFL-CIO headquarters has been handled by staffers George Taylor and Sheldon Samuels. Perhaps their greatest achievement has been to prevent the legislative weakening of the OSHA law, under tremendous pressure from business. The AFL-CIO has also sponsored lawsuits to strengthen the asbestos and benzene standards, and to coordinate new standards activity. Samuels also produces a quarterly newsletter which goes out to thousands of unionists. But only now, after years of discussion and proposals, has the long-awaited AFL-CIO center for health and safety begun to get off the ground.[111]

As we have seen, progress in health and safety has varied according to industry. Coalminers have been in a strong position to press their demands. Most coal is produced in unionized mines owned by large companies, and the demand for coal is strong. The public is sympathetic to the obvious hazards of the job, and the workers have repeatedly demonstrated their capacity to shut down the mines. Even so, union headquarters began to take strong stands only when the old leadership had been replaced. Even now the national union has retreated on the right to strike over health and safety. The asbestos workers' union, faced with a hazard which kills almost half of their members, has had some success in substituting other materials for asbestos, and in helping educate the public to fear asbestos. In California a group of naval shipyard workers has begun a campaign to remove asbestos from the ships, and is lobbying for the passage of a "white lung law," to compensate en masse the victims of asbestos and their families. So far the navy's attitude seems to be typified by the response of Captain George M. Lawton, a physician in the U.S. Navy and deputy director of the Occupational and Preventive Medicine Division. When asked if the navy was morally responsible for the working conditions of military contractors he replied: "If I order an automobile and the way they make automobiles is to throw people into a furnace, I am not responsible for that."[112] At Mare Island Naval Shipyard in Vallejo, California, the Metal Trades Council has worked closely with physician Phillip Polakoff to examine hundreds of its members for signs of asbestos disease. The California State Federation of Labor publicly backed the council's call for an end to the use of asbestos, medical screening exams for the workers, and adequate compensation for the disabled and the survivors.[113] To help deal with these and similar problems the State Federation and dozens of local unions have given their moral and financial backing to Polakoff's Western Institute for Occupational/Environmental Sciences at Herrick Hospital in Berkeley. In the spring of 1978 the Western Institute helped trigger a national screening program for shipyard workers.

Unions such as the Oil, Chemical, and Atomic Workers

(OCAW), the United Electrical Workers (UE), and the Amalgamated Clothing and Textile Workers Union (ACTWU), which are relatively weak in their industries, have used health and safety to organize new workers to strengthen ties with existing members. The ACTWU has given the Brown Lung Associations in the cotton industry both moral and financial support, and the OCAW has worked closely with environmentalists and other activists on a wide variety of issues to help overcome its relatively weak position in the petrochemical industry. Both the ACTWU and the UE have many members in small, competitive shops, where a long strike or large benefit increase could drive employers out of business. The OCAW deals largely with large companies, but has managed to organize only a portion of the highly automated petrochemical business. During the 1973 Shell strike the OCAW could not halt production, because supervisory personnel was able to keep the refineries running.

Health and safety advances in the automobile and steel industries have been mostly in the largest shops. As a union with an activist tradition, the UAW has trained hundreds of health and safety representatives from the Big Three auto companies. The Steelworkers' union, by contrast, has preferred to operate through OSHA, and at first shied away from systematic training of the members in health and safety. Finally, the construction trades, disciplined by extremely high unemployment rates, are fearful of doing anything that might raise employers' costs. The Teamsters Union, on the national level, has done almost nothing for the members' work safety. It is still legal for over-the-road truckers to log as many as seventy hours a week on the road.

But an analysis of the role of the unions has to discuss their function in the society as a whole. In the capitalist United States unions are tolerated, but definitely confined to the middle levels of power—more inclined to react rather than to lead. The unions cannot match the large corporations in their ability to buy talent to defend their interests, and they are usually subject to chronic internal dissension. Society at large is always potentially hostile, and over the years it has become more and more difficult for U.S. labor unions to organize new workers or keep those they have. Well-known labor commentator Sydney Lens estimates that the proportion of "organizable" workers (excluding supervisors, executives, and the self-employed) who are union members has dropped from 42 to 32 percent in the last two decades. Moreover, laws originally supposed to protect the workers' right to join unions are often used to impede recruitment. As the organized proportion of the workforce has declined, union power in Washington has waned; recently both the White House and the Congress have been lukewarm or hostile to organized labor's attempts to make organizing easier, even though for the first time in eight years both are controlled by Democrats that unions worked to elect.[114]

Within this context the passage of the OSHA act was something of an anomaly. Much of the momentum for its passage came from a few labor leaders who believed strongly in the issue. The public interest groups were very active in the lobbying. Some believe that Congress wanted to give labor its own environmental legislation, since "quality of life" issues were fashionable in 1970. The Steelworkers and the OCAW worked very hard for the OSHA law, and received a great deal of support from the national AFL-CIO. But even the Steelworkers have been surprised at the "profound impact of the law, which has gone far beyond our expectation at the time."[115] Judging from their initial response to the law, most national union leaders failed to realize the unsettling effect health and safety demands would have at the local level (see chapter 1). With some exceptions, most unions have made few attempts to forge rank-and-file alliances with other unions and the public over work environment issues. The upper echelons have ignored or discouraged groups like CACOSH, which organizes locally on health and safety. Labor's strong mass solidarity of the 1930s is gone, along with most of the leftists within the unions who incessantly preached working-class unity. Organizing new workers is nearly impossible without this cooperation, and many unorganized workers have turned cynical about the desire and ability of unions to protect them.

Unions in the United States have shown little interest in European labor's campaigns for more decision-making power over investment and the organization of production. They typically unite with business to claim that workers have no need to fight for a better environment and believe that more environmental protection takes away jobs. Jack Mundey, leader of an Australian construction union which halted $4 billion worth of projects that threatened public parks and housing, and who convinced the Australian Council of Trades Unions to ban the handling and the export of uranium, is anathema to American labor leaders.[116] In the United States union leadership is resolutely reluctant, in the words of a UAW vice-president, to challenge management's "sole responsibility"

> to determine the products to be manufactured, the location of plants, the schedules of production, the methods, processes and means of manufacture, and the administrative decisions governing finance, marketing, purchasing and the like.[117]

The health and safety question provides a powerful tool to demystify the corporation's use of science and its technical jargon to hide and justify dangerous conditions. In some factories, on health and safety matters, the union safety representatives have acquired a shadowy authority which rivals management's, and bright new leaders have used health and safety to build fighting reputations within the union, when other channels seemed closed to them. So far these protests have yet to crystallize into positive programs for workers' control of production. Hardly anyone in the unions even thinks in these terms. But as workers learn more, authority disperses downward, and both management and conservative unions find it harder to keep them under control.

|6|
The Future Politics
of Working Conditions

Occupational Health and the Economy

No one would deny that workers in the United States live better than they did in 1900: according to Census Bureau figures they earn more, work less, die later, and live in bigger houses.[1] More people spend more time in school than ever before, and union membership has risen from 3 to 24 percent of the labor force.[2] When no jobs are available, or when workers are too old or sick to work, they can now try for unemployment, social security, or welfare, all programs that were won through labor pressure. A high level of unemployment is permanent, but hardly anybody starves anymore.

Yet the relative positions of labor and capital remain unchanged since the turn of the century. A small minority owns most of the wealth and runs the country, and generally determines the health conditions of labor. Despite the organization of mass industrial unions, workers and the poor have been hard-pressed to maintain their share of the national income, let alone to control the hazards of the workplace.[3] As the world economic crisis deepens, inflation increases and profits fall, and unemployment rates have been allowed to go up in a deliberate attempt to cool labor's wage demands. As usual, the impact has been felt most heavily by women, minorities, and youth. Management has fought the unions to a standstill since World War II through repressive use of the legal apparatus, the constant introduction of labor-saving techniques, and through its freedom to move production to the South or abroad where unions are weak, labor is cheap, and social costs are almost nonexistent.[4]

Today U.S. capitalism is dominated more and more by gigantic companies and banks which plan their activities on a global rather than a national scale. They determine the direction of economic development largely through internally generated investment capital and friendly investment banks. In the United States, Western Europe, and Japan, with their relatively high wages, new factories usually use tremendous amounts of energy and capital and create very few jobs. The extraordinarily high labor productivity of the multinational corporations makes it possible for them to pay high wages, which sap organized labor's militance on other issues. Because they manage prices worldwide and operate in close coordination with the banks, these corporations are largely immune to attempts by national governments to regulate their activities. Indeed, through forums such as the Trilateral Commission, the policies of individual governments are coordinated to the needs of international monopoly capital.[5]

New Productive Techniques and the Work Environment

Petroleum, petrochemicals, and automobiles have replaced steel and railroads as the leading sectors in heavy industry, as compared to the turn of the century. Computers and other machines and systems have transformed work in offices as well. Jobs have been "rationalized" throughout the economy, leaving workers with more of less to do than ever before. These changes have had profound and often little understood consequences for working conditions. If the characteristic maladies of U.S. Steel in 1905 were accidents, tuberculosis, and exhaustion, the quintessential ills of the Age of Exxon and IBM would have to be stress and heart disease and cancer and boredom.[6] In the office the rationalization of "word-processing" has forced workers to invent new refinements in the art of staying awake. To speed up office work, management has employed the same basic time-and-motion techniques first developed for the assembly line.The Systems and Procedures Association of America has now published its techniques first developed for the assembly line. The Systems and Procedures Association of America has now published its useful research on the time it takes office workers to do what they do. "Chair activity," for example, has been broken down as follows:

	Minutes
Get up from chair	.033
Sit down in chair	.033
Turn in swivel chair	.009
Move in chair to adjoining desk or file (4 ft. maximum)	.050

Similar developments are occurring throughout the economy.[7]

The Problem of Stress

Capitalist enterprises, with their need to maximize profits and minimize production costs, have an inherently disruptive effect on people's lives. Marx argued that the "crippling of body and mind" was inseparable from the division of labor, which "attacks the individual at the very roots of his life." Periodic unemployment makes it difficult for workers and their families to plan their lives together, and subjects families to endless fights over money. Factory closings destroy communities by forcing people to migrate in search of other jobs. As technological innovations occur, whole classes of skills become unsaleable on the job market. The affected workers are coerced into learning new jobs, often at great personal cost in prestige and income, particularly if they are not unionized. When all these sources of stress are combined with speedup and the constant introduction of new chemical, physical, and electronic hazards, the effects are particularly deadly, as proven by the extra high death rates among black workers.[8]

So far the occupational health movement has concentrated on discrete, identifiable threats to life and limb—such as breathing asbestos or working at an unguarded machine— as if work will become healthy and safe when all machines are guarded and all poisons eliminated from the production process. But the evidence is mounting that the stress of living in an advanced capitalist society can kill and maim just as surely and unnecessarily as punching out parts at an un- guarded power press or spraying asbestos insulation. People suffer stress when facing a challenge which they cannot or do not know how to handle. In response, the body undergoes a process of physical and psychological arousal which pre- pares it for "fight or flight." In a few seconds or minutes the heart rate and blood pressure increase, blood is pumped to the brain and muscles from the stomach and skin, and energy-producing glucose and fatty acids are released into the blood stream. Of course, occasional stress is unavoidable and necessary for dealing with grave crises, and such stress can drive some people to new creative heights. Yet through processes only dimly understood, people who experience prolonged stress have excess rates of heart attacks, high blood pressure, cancer, ulcers, and other diseases. Divorce, the death of a spouse, migration, unemployment, speedup, and harassment on the job by an unfriendly boss have all been cited as causes of stress and elevated death rates.[9] The famous government report, *Work in America*, conceded tremendous importance to the role of work in a person's well-being and lifespan:

> Satisfaction with work appears to be the best predictor of lon- gevity—better than known medical or genetic factors—and various aspects of work account for much, if not most, of the factors associated with heart disease. Dull and demeaning work, work over which the worker has little or no control, as well as other poor features of work also contribute to an assortment of mental

health problems. But we find that work can be transformed into a singularly powerful source of psychological and physical rewards. . . . From the point of view of public policy, workers and society are bearing medical costs that have their genesis in the workplace, and which could be avoided through preventive measures.[10]

It is clear that the occupational movement by itself will be unable to directly attack the roots of stress, embedded as they are in the capitalist mode of production. But the movement will certainly be able to help by exposing and clarifying the connections between the built-in insecurities of life under advanced monopoly capitalism and stress, which conventional medicine usually blames on people's unhealthy lifestyles. Even the most effective antidotes to stress, such as transcendental meditation, are difficult to practice consistently because of the pressures of daily life. Most people cannot "slow down" and "take it easy" because they will lose their jobs if they do so.

The occasional experiments under capitalism in increasing worker participation in decision-making have proven successful in making work more pleasant and have relieved some of the stress and boredom for the worker. But even though they raise labor productivity in many cases, such experiments have been rare, because management is afraid that workers, emboldened and exhilarated by their ability to run the shop without bosses, might start trying to gain control of profits and investments as well. Then there would be nobody for the bosses to boss. Carried to its logical extreme, worker control would mean the end of capitalism. Thus, most experiments in worker participation have been management-sponsored ventures which set very clear limits to what workers may and may not decide. Many are no more than clever schemes to slip speedup and piecework in through the back door, and workers and unions are justifiably suspicious. Union leaders, in any case, have a hard time imagining a role for themselves in factories without traditional hierarchies and bosses. And when, as in Italy, Sweden, and West Germany, workers and their unions have demanded a major voice in running the enterprise, management opposition has been forceful and has succeeded in deflecting many of the demands.[11]

Certainly there is no way to make every job personally satisfying. A friend of mine who works on cars for his money and spends the rest of his time trying to learn photography said that "nobody should have to work more than three days a week." Madaline Jaundzems, a cabin attendant for Pan Am, believes that "everybody should make the same pay." There is a certain amount of dirty work to be done in any society. Perhaps both the work and the income from productive and highly paid but unpleasant jobs should be spread around more by shortening the work week and the work year. Some carpenters' union locals, for example, forbid overtime unless all members who want to work are working.[12] To partially free themselves from the weekly grind, some workers should be allowed to work long hours for six months, then go on leave for six months at no pay, but without losing their jobs or seniority. President Douglas Fraser of the United Auto Workers believes the four-day work week is inevitable:

I remember the old line: "If you don't come in on Sunday, don't come in on Monday." Moving to a reduced workweek is going to become a necessity in order to create enough new employment opportunities in coming years. The only question is how fast we can achieve it.[13]

Today, only students, teachers, scholars, and the wealthy are deemed worthy and able to benefit from so much time for themselves. On most routine but well-paying industrial jobs, workers who quit every few months lose all their accrued seniority and the benefits which go with it—it is the boss who decides when they work and when they do not. Charles Levinson, secretary-general of the International Chemical Workers Federation, despises most work in his industry:

We must have new concepts of work. I keep telling people, much to their amazement, one of the reasons I have been a trade union-ist all my life is that I never liked work, I think that it does very little good for you. I have learned in my own life that if you work hard you very seldom have time to make money and you certainly do not have time to develop the type of capacity that everybody is preaching is what the human entity should really be aspiring to. Most of it is an insult.[14]

International Competition and the Degradation of Working Conditions

In industries that make it impossible to substitute huge inputs of capital and energy for labor, or where U.S. occupational health and environmental standards are judged to be too expensive, hazardous work can be farmed out to labor-intensive assembly operations in poor countries dependent on international capital. A McGraw-Hill survey has shown that U.S.-based multinationals can get away with spending less than half on pollution control overseas than what they would spend on comparable new plants in the United States.[15]

For example, by 1973 Amatex, an asbestos textile manufacturer, had shut down its new headquarters plant in Norristown, Pennsylvania, and moved all of its production to the Mexican border towns of Ciudad Juarez and Agua Prieta. The asbestos fiber is now shipped from Quebec to Mexico, where workers are paid the minimum wage of about $5 per day. In 1977 Dr. William Johnson, an industrial health specialist, visited the plant in Agua Prieta with a reporter from the *Arizona Daily Star*. The reporter wrote that "asbestos waste clings to the fence that encloses the plant and is strewn across the dirt road behind the plant where children walk to school." He added, "Inside, machinery that weaves yarn into industrial fabric is caked with asbestos waste and the floor is covered with debris. Workers in part of the factory do not wear respirators. . . ." When the *Daily Star* article was printed in Spanish in the local Agua Prieta newspaper the plant workers called for an investigation by the Sonora State health authorities. As a result, the workers now wear uniforms to cover up their street clothes so that they bring home less asbestos to their families. Because of the scarcity of jobs none of the workers has quit, however, and the local union, not known for its independence, threatened to fire workers who continued to complain. Now the company cleans up the asbestos threads clinging to the outside of the building and the fence, but there are still clumps of asbestos hanging from bushes and roads near the plant.

Mexico, like most Third World countries, has no specific regulations regarding asbestos. Some general regulations on work health require a company to warn workers of health hazards and to protect them where necessary. The fine for violation of these regulations comes to about $45, or $90 for failing to take corrective action within a specified time.[16]

In Brazil, since the U.S.-backed military takeover in 1964, American and other foreign corporations have tightened their grip on the dynamic sectors of the Brazilian economy. Since there are almost no environmental restrictions, hazardous industries feel free to pollute the air and water and poison the workers. Asbestos production has skyrocketed, increasing from 3,700 tons in 1956 to 22,000 tons in 1971. By 1973 it had doubled, to 44,000 tons.[17] Independent unions have been destroyed or driven underground and giants such as Ford, General Motors, and Volkswagen often work their employees twelve hours per day, six days a week. Take-home pay has fallen by a half since the military takeover.[18] Although prices are as high as they are in the United States, most workers earn less than $100 per month. Moreover, employment in the automobile industry has become a game of musical chairs: people are fired routinely, before they can build up rights to pensions and other benefits, only to be rehired at base rates by another auto company. Working conditions are even worse for the tens of thousands of workers at small firms which supply parts to the automobile giants.[19] Not surprisingly, work accident rates are extremely high and have even attracted the attention of Brazil's president, General Geisel. According to figures from Brazil's National Institute for Social Insurance (INPS) for its 13 million subscribers (a quarter of the workforce), almost 4,000 workers were killed by accidents on the job in 1975.

Mimicking U.S. practice, private industry and the government have mounted a huge publicity campaign against work accidents. The government also modified the compensation law. Now employers rather than the INPS must pay for the first fifteen days of medical and workers' compensation benefits. As a consequence, fewer injured workers get medical care and more injuries go unreported to INPS. And predictably, the officially reported casualty rates have begun to drop.[20] Clearly, U.S. labor could help itself by uniting with overseas unions to struggle for better conditions.

However, the U.S. labor movement has largely cut itself off from the rest of world labor by its adamant refusal to deal with socialists and communists abroad. For years the AFL-CIO prevented communist unionists from visiting the United States; the ban was lifted only in 1977.[21] In addition, the AFL-CIO has actively allied itself with U.S. corporations, the State Department, and the CIA in policies designed to infiltrate and crush independent labor movements in Latin America and throughout the world. It also has helped build dual union structures which reject revolutionary or anti-imperialist policies directed against national domination by U.S. capital. The AFL-CIO has withdrawn from the International Confederation of Free Trade Unions (ICFTU), created with CIA help after World War II to fight communist and other leftist unions. George Meany, the aging president of the AFL-CIO, reportedly believes that the ICFTU has become "soft on communism" because it now favors co-operation, even with communist-led unions, against threats posed by multinational corporations.[22]

The American Institute for Free Labor Development (AIFLD), financed mostly by the CIA and backed by the AFL-CIO, claims to have trained a quarter of a million unionists at its Fort Royal, Virginia, training base and at other sites in Latin America.[23] These trainees have been instrumental in blocking more radical union efforts to effect social or political change in their native countries. In 1963, for example, the AIFLD coached a special group of 33 Brazilian union members in preparation for the military insurrection directed against the constitutionally elected regime of President "Jango" Goulart. When most Brazilian unions declared a general strike to try to block the military takeover, some of these trainees organized to break the strike and kept vital communications open to facilitate troop movements. Similarly, AIFLD actions in Chile helped make possible the overthrow of the Popular Unity government of Salvador Allende in 1973. Since then the military junta has unsuccessfully tried to abolish independent trade unionism in Chile.[24]

However, a few U.S. unions, such as the International Longshoremen's and Warehousemen's Union (ILWU), have always maintained their foreign ties, and the number is increasing. The United Auto Workers and the Machinists are active in the International Metalworkers Federation and lobbied to maintain United States membership in the International Labor Organization (ILO) in Geneva despite opposition from the AFL-CIO.

In occupational health, international cooperation is particularly important to fight international business dominance in the field of health and safety standards. Many European countries, including West Germany and Sweden, have borrowed most of their norms for permissible exposure to airborne poisons directly from OSHA or its private predecessor, the American Conference of Governmental Industrial Hygienists (ACGIH), with little critical examination. The ACGIH created the "threshold limit value" (TLV) concept in the late 1940s, and OSHA adopted these values as legally enforceable standards. As such, standard-setting abroad in occupational health is heavily influenced by the U.S. example.[25]

International union interest in joint efforts on occupational health began in 1973, triggered by findings linking vinyl chloride with liver cancer. Such links first were reported by Cesare Maltoni, an Italian scientist from Bologna. The Rubberworkers and Oil, Chemical, and Atomic Workers from the United States played a key role when the International Chemical Workers Federation (ICF) called a conference with representatives from West Germany, Britain, Sweden, France, and Japan to discuss a common approach to health and safety problems. Charles Levinson, secretary-general of the

ICF, forcefully argued for international union action on health and safety:

> There is no such thing as a national health hazard. It enjoys instant internationalization. The technologies today are applied globally. The same gases, the same particulates, the same dusts are found in factories of the same companies all around the world, and there is no distinction made between the nationalities or the place that they're used in, nor whether they are developed or underdeveloped countries. We believe, therefore, that it will be incumbent upon us, in co-operation with the best scientific intelligence available, to build new channels of communication, new forces of confrontation in order to make certain that this new hazard of ours does not remain a critical and threatening reality for the workers we represent around the world.[26]

The Meaning of Recent Developments

Because policy in both health and safety and workers' compensation was defined and executed by the compensation-safety apparatus, which combined funding, research, and information in the field, the first priority of those attempting to change policy was to attack the research and doctrines of the apparatus. New studies have shown that the accident rate is five or ten times what people had been led to believe, and that millions of workers are exposed to (usually unidentified) poisons on the job. It is now accepted that far more people are killed by slow-developing occupational diseases than by work accidents (see chapter 2).[27]

Despite all the official commissions and reports, workers' compensation is still a swindle—it is still predominantly a privately owned and state-regulated administrative system where most of the premium money ends up with the insurance carriers and a locusts' army of physicians, claims adjusters, inspectors, insurance agents, and lawyers. The major increases in benefits since the middle 1960s have been for temporary injuries rather than for much more costly permanent disability such as back injuries and respiratory diseases. In California and Illinois the insurance carriers and industry are already trying to roll back recent laws and rulings which have made it easier for injured workers to collect benefits for permanent disabilities.[28]

Rather than guaranteeing to workers the right to act directly in their own interests, the OSHA and Coal Mine Health and Safety laws allowed them the right to invoke the intervention of the government in their own behalf. But the OSHA administration was set up under the anti-labor administrations of Nixon and Ford, and like other agencies purporting to check corporate abuses, OSHA has generally been a captive to business, unless its hand is forced. The valiant efforts of Carter appointee Dr. Eula Bingham to turn OSHA around have been hampered by a virtual budgetary freeze, lack of support in Congress and on the highest levels of the Carter administration. The new OSHA regulation requiring employers to pay for workers' "lost-time" at work during OSHA inspections is being held up by litigation, and the administration's top economic advisor, Charles Schulze, has intervened against the advice of Eula Bingham to delay the implementation of a new cotton dust standard. In 1978 the OSHA law suffered a crippling blow to its ability to inspect without warning. The U.S. Supreme Court ruled that employers under inspection have the right to bar OSHA compliance officers who arrive at an inspection without a court-issued warrant authorizing a search. For employers with something to hide, the days of surprise inspections are over. The legal expenses for this campaign were raised largely by a group called "STOP-OSHA," spawned by the American Conservative Union, which raised over $200,000 by mail subscription of businesses which had been inspected by OSHA.[29]

In the immediate future the rising cost of energy is likely to have a harmful effect on working conditions, although little thought has been given to the problem. Factory owners, to make up for money lost to higher fuel prices, are forced to make their employees work harder and faster to preserve their profit margins. There is a tendency to cut back on nonessential maintenance and cleanup which benefit workers but do not raise labor productivity. Workshops are swept out less often, and the brakes on the forklifts are more likely to fail. Unless there is a breakdown, loose machine parts are ignored and create a terrific clatter. Fans and other protective devices which use a lot of power become relatively more expensive and cut more deeply into profit margins. When they finally wear out they are left to sit and rust. While workers could benefit from less exposure to airborne poisons, mechanization, ventilation, and enclosure become too expensive for management as the cost of fuel rises. Thus the effect of high-priced power may be to intensify already existing contradictions between productivity and workers' health, and to rule out the possibility of piecemeal solutions.

For products such as asbestos, the only reasonable answer may be a total prohibition on nonessential uses. This would be politically most difficult in countries such as the United States and Canada (where the asbestos business is large and profitable and where thousands of workers depend on it for their livelihood), and might well be a test of the seriousness of commitment of the socialist nations (whose asbestos production is skyrocketing) to protecting their workers. The beginnings of a ban are there. Growth in asbestos use is slowing in North America. The U.S. Navy now prohibits the use of asbestos in new ships, and precautions are strict when asbestos is encountered during refitting in naval shipyards.[30] In Birmingham, England, the Green Ban Action Committee, a group of environmentalists and unionists who refuse to work on "socially and environmentally harmful projects" points out:

> It is almost impossible to enforce [protective] standards on thousands of building sites, where asbestos board is sawn up, pipes lagged, and ceilings insulated. It is even more difficult for such workers to sue for compensation if they get asbestos diseases. If they have worked for a lot of companies, and on many different sites, who will they sue—how can they trace witnesses? *That is why it is even more important for construction workers to refuse to work with any asbestos in the first place.*[31]

Tasks for the Future

The most important tasks for the occupational health movement in the near future should be the preservation of legal rights already won and the building of new institutions to support the struggles of workers in the shops. The experience of the seventies has shown that if the issues are not dealt with by workers on the shop floor, they will not be dealt with at all. But experience has also shown that in order for workers to deal with the less obvious, long-term threats to health, they must have support from people knowledgeable about the science and tactics of occupational health. Without a distinct occupational health movement, the struggle will not advance. In addition, the movement in the United States must continue to build links with occupational health activists and workers' movements in other countries, in order to exchange experiences and build cooperative relationships on common problems. Following is a list of suggested strategies for action in the next five years:

1. Workers on the shop floor should move to create independent *"union-controlled* Safety and Health Committees in plants."[32] Funding to pay for training and education and for "lost time" from work should be demanded of the company (as the United Auto Workers and the Oil, Chemical, and Atomic Workers have done) or arranged by taxation of workers' compensation attorneys. People on these committees should be able to enforce demands on the company, and health and safety should be strike issues. In many cases these local committees will have to begin informally, against the wishes of the company and the upper levels of the union, but such committees are indispensable if local needs are to be served.

2. Union locals should band together with other union locals to form regional committees on safety and health ("COSH" groups). The COSH groups have sponsored a variety of programs, and have been tremendously helpful in breaking down the sense of isolation and impotence which safety committee members often feel in their individual

plants. The COSH groups have also been invaluable in monitoring the performance of OSHA, and in lobbying for stronger workers' compensation laws; and they have begun to cooperate with each other on a formal basis. In effect, they constitute the natural union counterparts to the management-controlled safety councils around the country.[32]

The Chicago and Philadelphia COSH groups have been the most successful. Both cities have large concentrations of unionized industrial workers. By contrast, in Pittsburgh and Detroit, headquarters of the steel and auto workers unions, such groups have been unsuccessful in getting off the ground, partly because of a lack of support from national unions (who saw them as competing power centers), partly because of a proliferation of competing leftist factions in those cities.

In the San Francisco Bay area, the Bay Area Committee on Occupational Safety and Health (BACOSH) fell apart, despite widespread interest from local unionists, because the intellectuals and health professionals who founded and dominated the group wasted their time in long arguments over the "political direction" of the organization. Workers who had traveled up to an hour to attend the meetings, after work, in search of practical assistance, soon drifted away from the organization. In this case many of the technical experts failed to perceive that the only way they could merit the confidence of workers and unions was to serve their immediate needs. The leadership of BACOSH was so mistrustful of unions that it didn't create a mechanism for local unions to join and pay dues.[33]

Two of the most active supporters of Pittsburgh's PACOSH quit the group because it lacked a base in the local unions, and because they disagreed with the "service" orientation of the group. Their criticism of PACOSH exemplifies the left criticism of many socialists who are dissatisfied with the COSH strategy:

> PACOSH has helped a number of workers to get technical information, good compensation lawyers, and medical help. But we have ended up being on a "service" organization—in a negative way. That is, we have led people to believe that PACOSH could

do things for them, rather than teaching and helping them do it themselves. We have been able to give out only "band-aid" solutions, to purely individual problems, instead of trying to find ways to attack the real roots of the problem. Technical information is helpful. But concentrating upon it too much leads to dependence on professionals—when in reality only workers can effect real change in the hazardous, sometimes fatal conditions on our jobs. Unless we begin to fight for the right to control our own working conditions, we will never be able to make our workplaces healthy and safe. Perhaps in the future there will be a positive role for an organization like PACOSH.[34]

3. The occupational health movement should campaign for the "right to know" the chemical names and hazards of all the substances they work with. In 1973 Local 3-562 of the Oil, Chemical, and Atomic Workers at a Ciba-Geigy plant in McIntosh, Alabama, won an arbitration decision which required the company to furnish the chemical names of all the hundreds of substances used in the plant, most of which had been known only by their trade names or code numbers. The arbitrator followed the union's reasoning that information about the chemicals was required for the purposes of informed collective bargaining. The OCAW's lead has been picked up by a coalition of COSH groups from Philadelphia, Chicago, Rhode Island, and Massachusetts. They have mounted a national petition campaign to persuade OSHA to issue regulations which force employers to reveal the chemical names and hazards of all substances used on the job, as required by Section 8(c)3 of the OSHA law.[35]

The "right to know" should also include the right to complete access by workers or their authorized physicians to company medical records and to records of their exposure to toxic substances. Unions should also be able to bring their own industrial hygienists into the plant if they deem it necessary. OCAW's 1974 Shell strike was fought over these issues.

4. The movement should fight to end the ghettoization of occupational medicine. Under the present system industrial doctors are a caste apart from normal medical practice. There are repeated accounts of company physicians who refuse to

inform workers about newly discovered hazards or potentially compensable diseases. By doing company occupational health research under university or medical school auspices, they lend legitimacy to attempts to whitewash or suppress known hazards.

In their attempts to hold down health care costs, monopoly sector corporations have now begun to experiment with extending the company medical system to cover, on a prepaid basis, all the health care needs of workers and their families.[36] If these "health maintenance organizations" (HMOs) were controlled by workers and the community, they would be useful, particularly in dangerous industries. But as industrial medicine is now organized, HMOs would constitute a new form of plantation medicine which would deprive participants of any independent voice in their health care. Under such a system the incentive to manipulate medical findings to control and fire workers would be considerably increased, and the workers would find it much more difficult to get an honest assessment of health problems on the job. The possibilities for abuse of psychiatry in dealing with what are now considered to be personal problems are frightening.

Workers should continue to fight company medicine through their own investigations, through third-party liability and medical malpractice suits, and through public accountability hearings whenever the inevitable horror stories come to light. The effect of these tactics has already shaken the self-confidence of company medicine, and created the conditions for company doctors, on occasion, to come forward with evidence of medical coverups.

5. The occupational health movement should call for international standards in health and safety, perhaps under the aegis of the International Labor Organization (ILO). After all, workers in similar industries face similar hazards throughout the world. Calling for international standards would focus attention on the practices of individual countries, and encourage more exchange and debate, from which everyone would profit. Soviet activities, for example, have been widely ignored in the United States, though the Soviet

Union has a long and proud tradition in safety and health. In 1923 the Soviet Union founded the first hospital in the world entirely devoted to the study and treatment of occupational diseases, which occupational health pioneer Dr. Alice Hamilton found "most impressive" in her visit to Moscow.[37] No such hospital exists to this day in the United States. The Soviets define their "maximum allowable concentrations" (MACs) for airborne contaminants so that no changes are observed in such relatively sensitive bodily areas as the nervous system. A typical test would be to find out at what concentration there is a slowing of conditioned reflexes or the ability to learn new material. As a result of their difference in philosophy, Soviet MACs are usually much lower than U.S. TLVs, often by a factor of 10 or more. How stringently the standards are enforced, however, is hard to determine.

Yet the Soviet Union faces many of the same problems with which the United States has yet to deal. Asbestos production has soared, and the petrochemical industry is growing much faster than the rest of the economy, often with the help of methods imported from the West. PPG Industries, notorious for the Tyler, Texas, asbestos atrocity, is building a giant petrochemical complex in the USSR for the production of vinyl chloride and other plastic resins.[38] "Maximum allowable concentrations" have been set for only a few hundred chemicals, and despite (or because of) the rapid growth in petrochemicals, many hazards are discovered only after production has begun, forcing the workers to serve as unwitting test animals. One Soviet scientist has written:

. . . in studies on workers in the chemical industry, it is usually very difficult to detect the agents mainly responsible for the pathological conditions produced because of the large number of factors that may be involved under industrial conditions. Nor is it possible to exclude the effect of exposure to chemicals in everyday life. . . . The main point is that the results of such research may not be available early enough to serve the prophylactic purposes of industrial hygiene.[39]

The Castleman report and the Seveso crisis in Italy, in which a whole region poisoned by a Swiss-owned factory had to be evacuated,[40] have shown that multinational corporations often transfer the production of poisons to less developed countries. Perhaps the International Chemical Workers Federation or some other international union body should create a study group to monitor the production and trade in poisons across international boundaries and suggest means for controlling or eliminating the problem.

6. The unions and their allies in the COSH groups should pressure the OSHA administration to provide more funds for direct training of workers in apprenticeship schools and from safety committees about work hazards. According to Morris Davis, LOHP director, most of OSHA's training money has been spent in company-dominated programs of the National Safety Council and at junior colleges.[41]

7. Unions should start to demand a role in investment, not only to insure healthy working conditions, but also to protect jobs and preserve neighborhoods. Often what is not built is as important as what is built. Local 1-5 of the OCAW in Richmond, California, questioned the construction of a giant chemical complex by Dow Chemical, arguing that the workers' health would be inadequately protected.[42] General Motors data about working conditions at the company's new and nonunion Delco-Remy battery plant in Fitzgerald, Georgia, disclosed extremely high lead levels, according to publicly available records.[43] One sample taken in 1975 was nine times the legal limit of 200 micrograms per cubic meter of air averaged over an eight-hour working day. Clearly, the best time to deal with potential problems is in the planning and design stage, rather than later. The problem is that unions cannot have a say before the plant is built, because there are no workers to represent.

The strategies discussed here are hardly revolutionary. They could lead, if successful, to the formation of a vigorous health and safety lobby for unionized workers, constant pressure to remedy dangerous conditions in unionized shops, a tripling in the total of workers' compensation benefits, and, in some industries, to a very substantial reduction in injuries

and occupational diseases. The effort devoted to rehabilitation of severely injured workers would be enhanced. But
most workers, who are unorganized, would find their working conditions little altered. Without a major shift to the left
in the political climate and a revitalization of the unions, the
occupational health movement will stall. Millions of workers
in small shops and in the South must be organized, and
unemployment eliminated, before workers will have the
strength to demand the programs they deserve to protect
their working conditions.

In the long run a strong, adequately funded, and uniform
national system of industrial inspection must be set up to
educate workers and discipline employers who refuse to
maintain decent working conditions. This was the unfulfilled
hope of the Occupational Safety and Health Act of 1970. A
national, publicly run system of workers' compensation
should be set up, perhaps under Social Security,[44] which
would eliminate the role of the private insurance companies;
and employers should be prohibited from firing workers who
are injured on the job. Medical care and rehabilitation for
sick and injured workers should be available under a national
health plan which makes health care free at the point of
delivery. Employers might be charged a "health and safety
tax" which would vary according to the casualties and excess
deaths caused by conditions at their workplaces. Since
statistics would be kept by a visible public organization under
some local control, it would be more difficult for companies
to bury their health and safety skeletons in the archives of
insurance companies and captive medical departments.
Replacing the private, state-level workers' compensation
system would finally end the pernicious ghettoization of
industrial medicine.

The distance we are from such a vigorous program to protect the lives of American workers on the job is one measure
of the struggles which lie ahead.

The problem of control of the process of production is the
explosive key to the issue of occupational health and safety.
A healthy work environment is incompatible with the survival of monopoly capitalism. The experiences so far in

health and safety show that progress becomes possible only through intensive education and bitter and sometimes bloody struggles. There is no question that the continued advance of the occupational health movement depends on the success of a popular movement to rechannel the country's energies to serve all the people.

Reference Notes

A word on U.S. Senate and House hearings

The U.S. Senate and the House of Representatives published thousands of pages of testimony before their relevant committees on the various occupational safety and health bills before Congress from 1968 through 1973. These hearings, which comprise some of the source material for this book, are listed below in chronological order, with the abbreviated titles used to refer to them in the footnotes. Testimony by individuals is listed under each individual's name in the footnotes.

1. *Occupational Health and Safety*, Hearings before the Select Subcommittee on Labor of the Committee on Education and Labor, House of Representatives, 90th Congress, 2nd session on H.R. 14816, 1968, one volume, about 800 pp., referred to as "1968 House hearings."
2. *Occupational Safety and Health Act of 1968*, Hearings before the Subcommittee on Labor of the Committee on Labor and Public Welfare, United States Senate, 90th Congress, 2nd session, on S. 2864, 1 vol., 10,004 pp., referred to as "1968 Senate hearings."
3. *Coal Mine Health and Safety*, Hearings before the Subcommittee on Labor and Public Welfare, United States Senate, 91st Congress, 1st session, on S. 355, S. 467,

S. 1094, S. 1178, S. 1300, and S. 1907, 5 vols., consecutively numbered, 1,786 pp., referred to as "1969 Senate coal mine hearings."
4. *Occupational Safety and Health Act of 1969*, Hearings before the Select Subcommittee on Labor of the Committee on Education and Labor, House of Representatives, 91st Congress, 1st session on H.R. 843, H.R. 3809, H.R. 4294, H.R. 13373, 2 vols., consecutively numbered, 1,594 pp., referred to as "1969 House hearings."
5. *Occupational Safety and Health Act, 1970*, Hearings before the Subcommittee on Labor of the Committee on Labor and Public Welfare, 91st Congress, 1st and 2nd sessions, on S. 2193 and S. 2788, 2 vols., consecutively numbered, 1,790 pp., referred to as "1970 Senate hearings."
6. *Occupational Safety and Health Act of 1970 (Oversight and Proposed Amendments)*, Hearings before the Select Subcommittee on Labor of the Committee on Education and Labor, House of Representatives, 92nd Congress, 2nd session; 1 vol., 754 pp., referred to as "1972 House oversight hearings."

Chapter 1

1. This account was put together from telephone and personal interviews with Marcos Vela and his wife in Antioch, California, in June 1975, February 1976, and August 1977, an article by D. B. Rubsamen in *Professional Liability Newsletter*, Berkeley, California, November, 1973.
2. James O'Connor, *The Fiscal Crisis of the State* (New York: St. Martin's Press, 1973), pp. 13–18.
3. The history of this period can be found in O'Connor, *The Fiscal Crisis of the State*; Gabriel Kolko, *Triumph of Conservatism* (New York: Free Press, 1963); James Weinstein, *The Corporate Ideal in the Liberal State* (Boston: Beacon Press, 1968); Philip Foner, *History of the Labor Movement in the United States*, vols. 2 and 3 (New York: International Publishers, 1955, 1964); and Ronald Radosh, "The Corporate Ideology of American Labor Leaders from Gompers to Hillman," *Studies on the Left* (November–December 1966): 66–96.

4. Richard Hofstadter, *The Age of Reform* (New York: Vintage Press, 1955), pp. 178, 179.
5. See, for example, C. V. Woodward, *The Strange Career of Jim Crow*, 3rd ed. (New York: Oxford University Press, 1975); W. E. B. DuBois, *The Souls of Black Folk* (New York: Fawcett Publications, 1961; originally published in 1903); and T. Rosengarten, *All God's Dangers: the Life of Nate Shaw* (New York: Avon Books, 1975).
6. Hofstadter, *Age of Reform*, pp. 178, 179.
7. Ibid., chapters 1–3.
8. Foner, *History of the Labor Movement*, vol. 2, pp. 376–87.
9. See James Weinstein, "The IWW and American Socialism," *Socialist Revolution* (September–October 1970): 3–42; and R. O. Boyer and H. W. Morais, *Labor's Untold Story* (New York: Cameron Press, 1955).
10. See Katherine Stone, "The Origins of Job Structures in the Steel Industry," *Review of Radical Political Economics* (Summer 1974): 61–97; and Upton Sinclair, *The Jungle* (1906; reprint ed., Cambridge, Mass.: Bentley, 1971).
11. Foner, *History of the Labor Movement*, vol. 3, p. 21.
12. F. L. Hoffman, "Industrial Accidents," *Bulletin of the Bureau of Labor* (September 1908): 418.
13. Mack Sennett, *King of Comedy* (New York: Doubleday, 1954), p. 19.
14. *Thirteenth Annual Report, 1895*, New York Bureau of Statistics of Labor, Part II, p. 5, as quoted in M. C. Cahill, *Shorter Hours* (New York: Columbia University Press, 1932), p. 125.
15. *Lochner v. New York*, 198 U.S. 45, 1905, as quoted in Cahill, *Shorter Hours*, p. 126.
16. Foner, *History of the Labor Movement*, vol. 3, pp. 20–24.
17. J. Sparge, *The Bitter Cry of the Children*, as quoted in Richard Hofstadter, *The Progressive Movement in the United States 1900–1915* (Englewood Cliffs, N.J.: Prentice-Hall, 1963), pp. 39–44.
18. W. Z. Foster, *American Trade Unionism* (New York: International Publishers, 1947), p. 12.
19. Stone, "Origins of Job Structures," pp. 64, 65; and John A. Fitch, *The Steel Workers* (New York: Arno Press, 1969; originally published in 1911).
20. See J. Brecher, *Strike!* (San Francisco: Straight Arrow Books, 1972), chapter 3.
21. Fitch, *Steel Workers*, p. 216.
22. D. Brody, *Steelworkers in America: The Nonunion Era* (Cambridge, Mass.: Harvard University Press, 1960), p. 100, as quoted in Joseph A. Page and Mary-Win O'Brien, *Bitter Wages* (New York: Grossman Publishers, 1973), p. 53.
23. P. S. Foner, "Comment," *Studies on the Left* (November–December 1966): 91–96.

200   *Death on the Job*

24. Fitch, *Steel Workers*.
25. T. Oliver, *Dangerous Trades* (London: J. Murray, 1902), p. 141, as quoted in Fitch, *Steel Workers*, chapter 7.
26. Fitch, *Steel Workers*, chapter 7.
27. D. Brody, *Steelworkers in America: The Nonunion Era*.
28. E. H. Gary, *Addresses and Statements*, vol. 4, January 21, 1919, as quoted in Stone, "Origins of Job Structures," p. 78.
29. On all of this see: W. Hard, "Making Steel and Killing Men," *Everybody's Magazine* (November 1907): 579-91; L. Palmer, "History of the Safety Movement," *The Annals of the American Academy of Political and Social Science* (hereafter *The Annals*) (January 1926); and G. A. Cowee, *Practical Safety Methods and Devices* (New York: Van Nostrand, 1916).
30. L. Chaney and H. S. Hanna, *The Safety Movement in the Iron and Steel Industry, 1907-1917*, Bulletin no. 234 (Washington, D.C.: Department of Labor Statistics, 1918), pp. 176, 177.
31. Ibid., p. 177.
32. G. M. Jensen, "The National Civic Federation" (Ph.D. diss., Princeton University, Princeton, N.J., 1956), pp. 325-30.
33. Stone, "Origins of Job Structures," p. 76.
34. Nicholas A. Ashford, *Crisis in the Workplace (Occupational Disease and Injury)* (Cambridge, Mass.: MIT Press, 1976), p. 47.
35. C. M. Destler, "The Opposition of American Businessmen to Social Control During the Gilded Age," *The Mississippi Valley Historical Review* (March 1953), p. 644.
36. See Weinstein, *The Corporate Ideal*; Jensen, "The National Civic Federation"; and William Domhoff, *The Higher Circles* (New York: Random House, 1970), chapter 6.
37. R. M. Easley, founder of the NCF, as quoted in Foner, *History of the Labor Movement*, vol. 2, p. 384; see also Weinstein, *The Corporate Ideal*, pp. 11, 12.
38. Foner, *History of the Labor Movement*, vol. 2, pp. 376-78.
39. Domhoff, *The Higher Circles*, p. 168.
40. A. Hamilton, *Exploring the Dangerous Trades* (Boston: Little, Brown and Company, 1943), pp. 8-11. Interestingly enough, one of her prime success stories, the National Lead Corporation, has recently attracted attention for poisoning a large number of workers in Indianapolis, then compounding the problem by "curing" them of the lead intoxication with a medicine that ruined their kidneys (*Spotlight on Health and Safety* [Washington, D.C.: Industrial Union Department, AFL-CIO, 1976]).
41. Page and O'Brien, *Bitter Wages*, chapters 3 and 4.
42. Weinstein, *The Corporate Ideal*, p. 117.
43. Jensen, "The National Civic Federation," p. 332.
44. Richard Hofstadter, *The American Political Tradition* (New York: Vintage, 1948), p. 223.

45. All of this is documented in ibid.; Hofstadter, *Age of Reform*, pp. 235-37, 246; and M. Edelman, *The Symbolic Uses of Politics* (Urbana, Ill.: University of Illinois Press, 1967), chapter 2.
46. Jensen, "The National Civic Federation," p. 132.
47. See Foner, *History of the Labor Movement*, vol. 2, p. 384 and vol. 3; Weinstein, *The Corporate Ideal* and "The IWW and American Socialism"; and Boyer and Morais, *Labor's Untold Story*.
48. Page and O'Brien, *Bitter Wages*, chapter 3.
49. R. Lubove, "Workmen's Compensation and the Prerogatives of Voluntarism," *Labor History*, pp. 259, 260.
50. On all of this see Weinstein, *The Corporate Ideal*, chapter 2, and Domhoff, *The Higher Circles*, chapter 6.
51. B. Marcus, "Defending the Rights of the Injured," in E. F. Cheit and M. S. Gordon, eds., *Occupational Disability and Public Policy* (New York: John Wiley and Sons, 1963), pp. 77-90.
52. D. N. Price, "Workers' Compensation: Coverage, Payments, and Costs, 1974," *Social Security Bulletin* (January 1976): 38-42; and C. D. Selby, "Studies of the Medical and Surgical Care of Industrial Workers," *Public Health Bulletin No. 99* (1919); as quoted in H. B. Selleck, *Occupational Health in America* (Detroit: Wayne State University Press, 1962), pp. 103, 106.
53. Palmer, "History of the Safety Movement."
54. See Close, "Safety in the Steel Industry," and Page and O'Brien, *Bitter Wages*, chapters 4 and 5.
55. See *The Annals* (January 1926), which devoted an entire edition to the problems of occupational safety.
56. C. B. Cox, "Putting Safety Across to the Worker," in ibid., pp. 191-96.
57. W. E. Powles and W. D. Ross, "Industrial and Occupational Psychiatry," in S. Arieti, ed., *American Handbook of Psychiatry* (New York: Basic Books, Inc., 1966), pp. 588-601. By contrast, Stone, "The Origins of Job Structure," examines details of the workplace, and Harry Braverman, *Labor and Monopoly Capital: The Degradation of Work in the Twentieth Century* (New York: Monthly Review Press, 1975), looks at the nature of work under monopoly capitalism.
58. W. H. Cameron, "Organizing for Safety Nationally," *The Annals* (January 1926): 27-32.
59. Ibid.
60. See S. P. Orth, *The Armies of Labor* (New Haven, Conn.: Yale University Press, 1919), pp. 30-35.
61. Brecher, *Strike!*, chapter 2.
62. Foner, *History of the Labor Movement*, vol. 2, pp. 376-87.
63. Weinstein, *The Corporate Ideal*, p. 43.
64. C. Eastman, *Work-Accidents and the Law* (New York: Russell Sage Foundation, 1910), pp. 286-90.

65. This issue is discussed by Weinstein, *The Corporate Ideal*, chapter 2 (New York); R. Kerstein and E. Goldenhersch, *Workmen's Compensation in Missouri* (St. Louis: Missouri Public Interest Research Group, 1972), photocopy (Missouri); M. Rosenblum, ed., *Compendium on Workmen's Compensation* (Washington, D.C.: National Commission on State Workmen's Compensation Laws, 1973), p. 272 (Ohio); and Price, "Workers' Compensation, 1974"; and " 'Get Mine' in Ohio," *Time*, August 23, 1976.
66. Page and O'Brien, *Bitter Wages*, pp. 59-63.
67. Ibid.
68. R. Scott, *Muscle and Blood* (New York: E. P. Dutton and Company, 1974), pp. 176-96.
69. Ibid., p. 179.
70. E. A. Balanoff, "History of the Black Community of Gary, Indiana, 1906-1940" (Ph.D. diss., University of Chicago), pp. 215-19.
71. S. Nowicki, "Back of the Yards," in *Rank and File*, ed. Alice and Staughton Lynd (Boston: Beacon Press, 1973), pp. 67-88.
72. Victoria M. Trasko, "Status of Occupational Health Programs in State and Local Governments, January 1969" in *Occupational Safety and Health Act, 1970*, hearings before the Subcommittee on Labor, Committee on Labor and Public Welfare, United States Senate, pp. 103-09.
73. H. K. Abrams, "Diatomaceous Earth Pneumoconiosis," *American Journal of Public Health* (May 1954): 592-99. While the problems with diatomaceous earth at Lompoc have been mostly resolved, new investigation has revealed that Johns-Manville has been exposing workers there to large amounts of asbestos dust for years, without informing them. As a result many have died (S. Tulledo, *Lompoc Record*, Lompoc, California, December 3, 1975, p. 1).
74. R. Ginnold, "Workmen's Compensation for Hearing Loss in Wisconsin," *Labor Law Journal* (November 1974): 682-97.
75. See "Bill Worthington, Lee Smith, and Ed Ryan" in *Rank and File*, ed. Alice and Staughton Lynd, pp. 285-96; H. W. Benson, "Labor Leaders, Intellectuals, and Freedom in the Unions," *Dissent* (Spring 1973): 206-16; and R. Diehl, "UMWA Reform Insurgency—A Recent History," *People's Appalachia* (Winter 1972-1973): 4, 5.
76. See Page and O'Brien, *Bitter Wages*, chapter 7.
77. Ibid., p. 144.
78. G. C. Guenther, letter. Republished in *Facts and Analysis*, Industrial Union Department, AFL-CIO, July 22, 1974.
79. All in Ashford, *Crisis in the Workplace*, pp. 258, 259, 260.
80. B. Cottine, with L. Birrel and R. Jennings, *Winning at the Occupational Safety and Health Review Commission* (Washington, D.C.: Health Research Group, 1975).
81. J. G. Hyatt, "U.S. Inspection Unit Finds Itself in Critical Crossfire," *Wall Street Journal*, August 20, 1974.

82. See Jane Brody, *New York Times*, February 16, 1974, and B. Cottine and R. R. Pancake, "In the Matter of the Economic Impact Study on the Proposed Permanent Standard on Occupational Exposure to Vinyl Chloride," *Comments* (Health Research Group, before the Occupational Safety and Health Administration, U.S. Department of Labor, September 6, 1974).
83. See, for example, J. B. Schoenberg and C. A. Mitchell, "Implementation of the Federal Asbestos Standard in Connecticut," *Journal of Occupational Medicine* (December 1974): 781–84.
84. E. Faltenmeyer, "Ever Increasing Affluence Is Less of a Sure Thing," *Fortune* (April 1975): 92ff.
85. L. Beman, "The Slow Road Back to Full Employment," *Fortune* (June 1975).
86. R. J. Barnet and R. E. Müller, *Global Reach* (New York: Simon and Schuster, 1974); and H. Magdoff, *The Age of Imperialism* (New York: Monthly Review Press, 1969).
87. See O'Connor, *The Fiscal Crisis of the State*; "Notes and Comments," *The New Yorker* (October 27, 1975): 31–33; and Notice in Boston Edison Company home electric bill explaining why the utility was switching to high-sulphur coal for power generation, February 1976.
88. See A. H. Raskin, "Nonunion Workers Are an Expanding Majority," *New York Times*, November 2, 1975, Section 3; and "Labor: Public Aggression, Private Cooperation," *New York Times*, August 31, 1975, Section 3.
89. On all of these recent trends see *First Annual McGraw-Hill Survey of Investment in Employee Safety and Health* (New York: McGraw-Hill, May 25, 1973); "Labor Letter," *Wall Street Journal*, October 12, 1976, p. 1; J. O. Hyatt, "Proposal Labor Agency Quit Safety Role Makes Political Health for Administration," *Wall Street Journal*, July 18, 1977; D. B. Kalis, "Workmen's Comp: Will Uncle Sam Deal Himself In?" *Occupational Hazards* (December 1975), pp. 35–38; "Illinois Workers' Comp Law—What Next?" *CACOSH Health and Safety News*, Chicago (July 1977); and Industrial Indemnity (a Crum and Forster Insurance Company), *Annual Injury (the Problem Can Be Solved)* (San Francisco, 1977).

Chapter 2

1. On the uses of information, see Harold L. Wilensky, *Organizational Intelligence* (New York: Basic Books, 1967); and Murray Edelman, *The Symbolic Uses of Politics* (Urbana, Ill.: University of Illinois Press, 1967; original clothbound, 1964), esp. chapter 7, "The Forms and Meanings of Political Language."
2. Richard J. Barnet and Ronald E. Müller, *Global Reach: The Power of the Multinational Corporations* (New York: Simon and Schuster, 1974), esp. chapter 10, "The Global Corporation and the Public Interest: The Managerial Dilemma of the Nation-State"; Ronald E. Müller, "The Political Economy of Global Corporations and National Stabilization Policy," prepared for The American Assembly, Columbia University, December 12-15, 1974, photocopy, 20 pp.; and Newton W. Lamson, "When the Auditor Gets Audited," *New York Times*, November 23, 1975, Section 3, pp. 1, 14.
3. John Rothchild, "Finding the Facts Bureaucrats Hide," *Washington Monthly* (January 1972): 15–27.
4. Frederick L. Hoffman, "Industrial Accidents," *Bulletin of the Bureau of Labor* 78 (September 1908): 417, 418.
5. Frederick L. Hoffman, "Industrial Accident Statistics," *Bulletin of the U.S. Bureau of Labor Statistics* (March 1915): 5, 6. Women were apparently included in this estimate.
6. For 1921 and 1919, respectively. See Lucien W. Chaney's *Statistics of Industrial Accidents in the United States to the End of 1927* (Bureau of Labor Statistics Bulletin no. 490, 1928). See also *Bulletin* nos. 339 (1923) and (425 (1927) for more Labor Department statistics on job injuries and accidental deaths.
7. *Accident Facts, 1972* (Chicago: National Safety Council), p. 25.
8. In *Accident Facts, 1975 Final Condensed Edition* (Chicago: National Safety Council, July 1975) there is noted a 6 percent decline of work accident deaths for 1974 over 1973, and *Accident Facts, 1972* records a decrease of 14 percent in the total number and 45 percent in the rate of work accidents between 1945 and 1971. Even if this were true, it would not necessarily be due to improved working conditions, but to improvements in the treatment of trauma. E. A. Vastyan, in "Civilian War Casualties and Medical Care in South Vietnam," *Annals of Internal Medicine* 74 (1971): 611–624, has pointed out that "Among [U.S.] combat casualties who reach medical aid, the death rate has fallen from 4.8% in World War II and 2.5% in Korea to but 1% in Vietnam." Ivan Illich, one of medicine's severest critics, has admitted, in *Medical Nemesis* (London: Calder & Boyars, 1975), that "more people survive trauma" in recent years (p. 20).
9. Vincent L. Tofany, President, National Safety Council, testimony at 1974 Senate hearings, "Occupational Safety and Health Act of

1970," July 22, 1974, "Supplement," p. 2, available from NSC, Chicago.

10. *Accident Facts, 1972*, p. 3. Data are for 1971 and come from the National Center for Health Statistics, based on death certificate counts.

11. Lillian Guralnick, in "Mortality by Occupation and Industry Among Men 20 to 64 Years of Age, United States 1950," *Vital Statistics—Special Reports*, 53, no. 2 (September 1962) (U.S. Public Health Service, Vital Statistics Division), found, based on death certificate information, that of 34,949 accidental deaths of men aged 20 to 64 with work experience, 8,297 died "while at work," and 26,652 died accidentally "not while at work and not stated." The unreliability of death certificate cause of death information is a rule-of-thumb among epidemiologists.

12. Tofany, "Supplement," p. 2.

13. See *Accident Facts, 1972*, p. 34, table entitled "Work Injuries and Compensation Payments, 1970," for 37 of 54 jurisdictions, which reported only 7,888 deaths for that year. How the NSC can put together death estimates for 1971 based on incomplete 1970 workers' compensation death totals is beyond my ken. For evidence on the absolute unreliability of workers' compensation data for statistical purposes see Monroe Berkowitz, *The Processing of Workmen's Compensation Cases* (Bureau of Labor Standards, Bulletin no. 310, 1967), pp. 16–18; and also his "Sources of Information About Workmen's Compensation Recipients," in *Supplemental Studies for the National Commission on State Workmen's Compensation Laws*, p. 109.

14. Interview by author with Robert J. Reszek at NSC headquarters in Chicago, February 8, 1973. President Tofany's testimony regarding the NSC's statistical techniques, with its vague references to unnamed experts and institutions, has all the scientific subtlety of a "scientifically proven" advertisement:

> In comparing these figures the Council's statisticians use the absolute numbers, the change indicated by comparable reports, and the rates per hundred thousand workers. Any estimates made must be consistent with the reports from the various sources and with trends from previous years. Any unusual experience is probed for that reason. . . . [Recently] NSC made a special effort to examine its estimating procedure even more carefully. Special detailed data for several large states and discussion with state vital statisticians were used to develop a work fatality estimate by another method. This additional estimate verified the accuracy of the NSC estimating procedure.

15. For 1922 and 1925 respectively; see *Bulletins of the Bureau of Labor Statistics*, nos. 339 (1923), 425 (1927), and 490 (1928).

16. In Bureau of Labor Statistics *Bulletin* no. 276.

17. The "USA Standard Method of Recording and Measuring Work Injury Experience—Z16.1," in 1970 Senate hearings, p. 1176.

18. McElroy was quoted in Charles Pearce, "Quality of Statistics on Work Injury Rates," a paper presented at the Interstate Conference on Labor Statistics, Knoxville, Tennessee, July 9, 1959, pp. 13, 14; cited in Joseph A. Page and Mary-Win O'Brien, *Bitter Wages* (New York: Grossman, 1973), pp. 161–165.

19. Ray Davidson, *Peril on the Job* (Washington, D.C.: Public Affairs Press, 1970), p. 153.

20. Jerome B. Gordon, Allan Akman, and Michael L. Brooke, *An Evaluation of the National Industrial Safety Statistics Program* (Delphic Systems and Research Corporation, Department of Labor Contract L-69-87, September 1970), chapter 9. Hereafter "Gordon Report."

21. Nicholas A. Ashford, *Crisis in the Workplace* (Cambridge, Mass.: MIT Press, 1976), p. 133n.

22. "Slightly Over One in Every Ten Private Nonfarm Workers Suffered Injury or Illness During 1972," *Daily Labor Report*, Economic Section (Bureau of National Affairs, January 21, 1974).

23. *What Every Employer Needs to Know About OSHA Recordkeeping*, Report 412 (revised) (Bureau of Labor Statistics, 1973), p. 2.

24. See "Doctoring the Facts," *Health Rights News* (March 1972), p. 11, and Vincent L. Tofany, "Supplement," p. 1.

25. David P. Discher, Goldy D. Kleinman, F. James Foster, *Pilot Study for Development of an Occupational Disease Surveillance Method* (NIOSH, May 1975), p. 54. Hereafter "Seattle Pilot Study."

26. Theodore Golonka, "Not as Simple as One, Two, Three," *Job Safety and Health* (August 1975): 23, 24.

27. William Domhoff, *The Higher Circles* (New York: Random House, 1970), esp. chapter 6; also Dr. Alice Hamilton, *Exploring the Dangerous Trades* (Boston: Little, Brown and Company, 1943), pp. 9–13.

28. Victoria M. Trasko, "Occupational Injury and Disease Reporting in the U.S.: A Status Report" (Cincinnati, Ohio: Bureau of Occupational Safety and Health, March 1971).

29. Andrew Kalmykow, Counsel, American Insurance Association, in 1970 Senate hearings, p. 349.

30. On this issue see Victoria M. Trasko, *Occupational Disease Reporting* (U.S. Public Health Service Publication no. 288, November 9, 1953); E. J. Baier, "What Occupational Health Problems Confront Us Today?" *Transactions of the 32nd Annual Meeting of the ACGIH*, p. 103. Arend Bouhuys and J. W. Peters in "Control of Environmental Lung Disease," *New England Journal of Medicine* (September 10, 1970), point out that "There are no systematic data on the prevalence of most occupational lung diseases. . . ." See also *Occupational Disease in California, 1968* (State of California, Department of Public Health, Bureau of Occupational Disease and Environmental Epidemiology), pp. 10–13, 24; and finally Harriet

L. Hardy, "Beryllium Poisoning—Lessons in Control of Man-Made Disease," *New England Journal of Medicine* (November 25, 1965): 1188-1200.

31. *Occupational Disease in California, 1968* reported 19,512 cases of occupational disease and 10,000 cases of "eye conditions due to toxic chemicals and chemical burns" for the 80 percent of the California workforce covered by compensation. Thus for all California workers one might project 36,900 cases of occupational disease. Since California is about ten percent of the U.S. workforce, we would project about 369,000 new cases of occupational disease annually in the U.S., based on the California workers' compensation definition.

32. Thomas F. Mancuso, *Medical Aspects* (U.S. Department of Labor, Interdepartmental Workers' Compensation Task Force [IWCTF], Conference on Occupational Diseases and Workers' Compensation, February 10-12, 1976), photocopy, p. 72.

33. Ibid., p. 73.

34. Richard Ginnold, "Workmen's Compensation for Hearing Loss in Wisconsin," *Labor Law Journal* (November 1974): 682-697; Rachel Scott's excellent account of industry's successful attempt to limit the compensability of silicosis after national attention was focused on the disease by the Gauley Bridge disaster in the early 1930s (*Muscle and Blood*, chapter entitled "Science—the Drab and Slut of Industrialism," pp. 174-203).

35. See both *Occupational Disease in California, 1968*, p. 24 and Berkov, "Evaluation of Occupational Disease Reporting in California." Management-oriented authors oppose attributing heart attacks and strokes to occupational "strain." See W. M. Gefafer, principal ed., *Occupational Diseases, A Guide to Their Recognition* (U.S. Public Health Service Publication no. 1097, 1964), esp. p. 5.

36. On January 26, 1973, at a lecture at the University of Illinois Medical School in Chicago, Dr. Joseph Wagoner of NIOSH estimated that 2,500 excess deaths per year were due to asbestos exposures in the United States. California, with one-tenth the national workforce, and most of the U.S. asbestos mining, might be expected to have at least 10 percent of the excess deaths annually, or 250. My estimate of 7 deaths annually reported by the workers' compensation system in California as caused by asbestos is high, because it sums all deaths due to "asbestosis" and all "malignant neoplasms." Figures are from *Occupational Disease in California, 1968*, Table 11, p. 24. Where we would expect 250 deaths annually attributable to asbestos disease, by stretching the data we can attribute 7 deaths as reported by the workers' compensation system, or 36 times below the estimated true incidence.

37. For a review of the efforts to win compensation status for these diseases, see Scott, *Muscle and Blood*, pp. 174-203 (silicosis); Ginnold, "Workmen's Compensation for Hearing Loss in Wisconsin"

208   *Death on the Job*

(hearing loss); and Mancuso, *Medical Aspects*, pp. 59–67 (cancer). Mancuso vehemently complains about the use of workers' smoking histories to deny the existence of work-related lung and other cancers. On brown lung see: Dr. I. E. Buff, 1970 Senate hearings, pp. 576–84, for a personal account of conditions in the Southern textile mills; also literature from the Brown Lung Association, 312 Pall Mall Street, Columbia, South Carolina, including article from *The State*, November 13, 1975, by Jan Stucker. The technical term for brown lung is "byssinosis."

38. H. W. Benson, "Labor Leaders, Intellectuals and Freedom in the Unions," *Dissent* (Spring 1973): 206–16; and Rick Diehl, "UMWA Reform Insurgency—A Recent History," *People's Appalachia* (Winter 1972–73): 4, 5.
39. In *Public Health Reports* (September 1958): 839–46.
40. See esp. p. 111 and p. 128, Table 5-1. The groups for which excess death rates were observed included underground metal miners, smelter workers, and uranium miners.
41. The American Cancer Society's publication '*71 Cancer Facts and Figures* estimates that 89,900 deaths were due to cancers of the lung, kidney, liver, and bladder. If even part of the over 900,000 deaths caused by heart and blood vessel disease are attributable to work exposure, then the figure of 100,000 deaths caused by work exposure is believable. See also, "Cancer on the Job—'Alarming' U.S. Study," *San Francisco Chronicle*, September 12, 1978, p. 1.
42. "Seattle Pilot Study." The institution conducting the study was the Department of Environmental Health, School of Public Health, University of Washington, Seattle, Washington.
43. Ibid., Table 12b and p. 56.
44. Missouri Division of Workmen's Compensation, *Missouri Workmen's Compensation Law* (42nd Annual Report, 1969).
45. Berkowitz, *The Processing of Workmen's Compensation Cases*, pp. 16–18. See also "Seattle Pilot Study," which reported a 16.1 percent "yes" answer to the question "Did you have a serious accident or injury on the job in the past 12 months? For example broken nose, brain concussion, deep cut requiring stitches, back injury, particle in eye, smashed finger?" (Table 9c). According to the "Gordon Report," chapter 9, about one worker in three has a "serious" but "non-disabling" injury every year.
46. Letter to author, April 2, 1971.
47. *Accident Facts, 1975, Final Condensed Edition* still does not mention occupational disease as a cause of workplace mortality.
48. Golonka, "Not as Simple as One, Two, Three," pp. 23–28. According to Goldy D. Kleinman of the University of Washington, Seattle, NIOSH has lost interest in the problem of occupational disease incidence (letter to author from Goldy D. Kleinman, December 15, 1975).

49. Tofany, "Supplement."
50. "Results of Survey of Occupational Injuries and Illnesses for 1974" (U.S. Department of Labor, Bureau of Labor Statistics, November 17, 1975), pp. 75, 647. Goldy D. Kleinman, who pointed this out to me, finds this "way off-beat" (letter to author, June 1, 1976).
51. A partial list of National Health Survey studies touching the problem of occupational injury is as follows: "Selected Health Characteristics by Occupation, United States—July 1961–June 1963," *Vital and Health Statistics from the National Health Survey*, National Center for Health Statistics, U.S. Public Health Service, Series 10, no. 21, August 1965), Table 22, p. 47; "Impairments Due to Injury by Class and Type of Accident, United States–July 1959–June 1961," Series 10, no. 6 (January 1964), Table 3; "Current Estimates from the Health Interview Survey, United States— July 1965–June 1966," Series 10, no. 57 (May 1967). For a short discussion of some of these findings, see author's Ph.D. dissertation, "Death on the Job: The Politics of Occupational Health in the United States" (Department of Political Science, Washington University, 1974), available on University Microfilms. The "Gordon Report," pp. 9–41, objected that the "National Household Surveys do not, because of the few numbers of work injury and re-related information developed, permit the collection of anything but national aggregate data." But this is precisely what is lacking at present! The "Gordon Report" also cites the limitations of the information available from the National Health Surveys: proxy respondents for those workers not available at interview time, and limitations on the accuracy of what workers (or other household members) report to the interviewers. I believe that such workers have less motivation to underestimate or lie than company reporters, and I believe the evidence in this chapter bears me out. Survey techniques can be designed to meet some of these objections.
    The "Gordon Report" also suggests using the Current Population Survey, Bureau of the Census, for occupational injury surveys. This survey develops monthly data on employment, unemployment, and other subjects dealing with workers. Monroe Berkowitz, in "Sources of Information about Workmen's Compensation Recipients," p. 121, also advocates the use of National Health Survey to get an accurate idea about the recipients of workers' compensation.
52. Robert D. Verhalen, "Injury Intelligence: Keying Product-Related Accidents," *Trial* (January–February 1972): 32, 33. The NEISS System still functions as this article describes it, according to Andrea Hricko, a former accident investigator for NEISS. For more information, write the Consumer Product Safety Commission, Washington, D.C. 20207. For an idea of how emergency room systems and hospitals could respond better to the needs of those

hurt on the job, send for "Occupational Health Services in the Hospital Clinics and Emergency Rooms," by Frank Goldsmith, Guide #2, Labor Safety and Health Institute, 381 Park Avenue South, New York, N.Y. 10016.

53. In the long run a national health service controlled by the public and the workers can be expected to wrest control of industrial medicine away from the companies. Then factory clinics and hospital emergency rooms will become reliable sources of workplace casualty statistics.
54. Goldy D. Kleinman, letter to author, December 15, 1975.
55. See, for example, Hardy, "Beryllium Poisoning—Lessons in the Control of Man-Made Diseases."
56. Mancuso, *Medical Aspects*, pp. 85–90.

Chapter 3

1. *Analysis of Workers' Compensation Laws*, 1976 ed. (Washington, D.C.: U.S. Chamber of Commerce), pp. 24, 25.
2. Since hardly any of the costs of occupational diseases are reimbursed by compensation (even though diseases are many times more deadly than accidents), the 10 percent estimate is probably a generous one.
3. "Labor Letter," *Wall Street Journal*, October 12, 1976, p. 1.
4. Daniel N. Price, "Workers' Compensation: Coverage, Payments, Costs, 1974," *Social Security Bulletin* (January 1976): 38–42.
5. Nicholas A. Ashford, *Crisis in the Workplace: Occupational Disease and Injury* (Cambridge, Mass.: MIT Press, 1976), pp. 260, 326.
6. Victoria M. Trasko, "Silicosis, A Continuing Problem," *Public Health Reports* (September 1958): 845.
7. See Willard W. Wirtz, Secretary of Labor, testimony with Esther Peterson, Assistant Secretary of Labor for Labor Standards, and Charles Donahue, Solicitor, Department of Labor, 1969 Senate hearings, p. 92; and Ralph Nader and Jerome B. Gordon, 1968 Senate hearings, p. 574.
8. See Jacob Clayman, Administrative Director, Industrial Union Department, AFL-CIO, 1970 Senate hearings, p. 422 (Ohio); Daniel M. Berman, "Occupational Diseases and Public Policy: the Case of Industrial Lead Poisoning in Missouri" (unpublished ms., photocopy) (Missouri); Anthony Mazzocchi, Legislative-Citizenship Director, Oil, Chemical, and Atomic Workers; 1968 House hearings, pp. 392, 393 (New Jersey); Daniel W. Hannan, president, United Steelworkers Local 1557, Clairton, Pennsylvania, 1970

Senate hearings, pp. 926, 946 (Pennsylvania); "Coroner's Verdict on the Death of Claudie B. Douglas," January 11, 1970, reproduced in 1970 Senate hearings p. 1587 (Indiana); and Urban Planning Aid, Inc., "The Enforcement of Occupational Safety and Health Laws in Massachusetts: A Critique" (Cambridge, Mass., 1971, mimeo).

9. For further information see Berman, "Occupational Diseases and Public Policy." George Flexsenhar was interviewed by the author in February 1970 in Herculaneum, Missouri.

10. Hannan, Senate testimony, 1970, pp. 925, 927, 936.

11. F. K. Barnako, Manager, Compensation and Safety Division, Bethlehem Steel Corporation, 1969 Senate hearings, p. 445.

12. Joseph A. Page and Mary-Win O'Brien, *Bitter Wages* (New York: Grossman, 1973), pp. 94–100, 157.

13. Material on insurance company programs can be found in the 1968 Senate hearings, including *Insurance Expense Exhibit (Countrywide)*, for the year ending December 31, 1966; compiled by the National Council on Compensation Insurance, 1967, reprinted in 1968 Senate hearings, p. 820; "Safety Activities of Compensation Insurance Companies," in 1968 Senate hearings, p. 839; and Frederick W. Deeg, safety engineer, American Mutual Insurance Alliance, 1968 House hearings, p. 438; see also Page and O'Brien, *Bitter Wages*, p. 157; and David Van Schaak, "The Part of the Casualty Insurance Company in Accident Prevention," *The Annals of the American Academy of Political and Social Science* (hereafter *The Annals*) (January 1926): 36–40.

14. S. Menshikov, *Millionaires and Managers* (Moscow: Progress Publishers, 1969), p. 146; also chapters 4 and 5.

15. Insurance company–bank interlocks do not violate antitrust laws; see "Case Against S.F. Bankers Ruled," *San Francisco Chronicle*, September 1976. On interlocks between the property-casualty insurance industry and the rest of finance capital, see *Disclosure of Corporate Ownership*, prepared by the Subcommittee on Intergovernmental Relations of the Committee on Governmental Operation, U.S. Senate, March 4, 1974, pp. 146, 157, 160, 333; G. William Domhoff's *Who Rules America?* (Englewood Cliffs, N.J.: Prentice-Hall, 1967), pp. 38, 53, 54, 164.

16. See *Best's Aggregates and Averages (Property-Liability)*, 1975 ed., pp. 1, 64, 123; and *Property and Casualty Insurance Companies— Their Role as Financial Intermediaries*, monograph prepared for the Commission on Money and Credit, the American Mutual Insurance Alliance, Association of Casualty and Surety Companies, and National Board of Fire Underwriters (Englewood Cliffs, N.J.: Prentice-Hall, 1962).

17. See Nader and Gordon, 1968 Senate hearings, p. 555; Morton Keller, *The Life Insurance Enterprise, 1885-1910* (Cambridge,

Mass.: Harvard University Press, Belknap Press, 1963), p. 211; and David Serber, "Regulating Reform: The Social Organization of Insurance Regulation," *The Insurgent Sociologist* (Spring 1975): 83–105.

18. David McCahan, *State Insurance in the United States* (Philadelphia: University of Pennsylvania Press, 1929), chapter 2.

19. G. Arthur Williams, *Insurance Arrangements Under Workmen's Compensation* (U.S. Department of Labor Bulletin No. 317, 1969), p. 109.

20. Jonathan C. Rose, Deputy Attorney General, antitrust division, U.S. Department of Justice, "Should Competition Be National Policy?" (address to the Institute on Regulation of the Insurance Industry, American Bar Association, New Orleans; text in *The National Underwriter, Property and Casualty Division* [18 June 1976]).

21. William Leslie, "Workmen's Compensation Insurance," *The Annals* (May 1932): 98–104.

22. Miles R. Drobisch, assistant actuary, California Inspection Rating Bureau, telephone conversation with author, September 16, 1976.

23. *Annual Report*, National Council on Compensation Insurance, March 4, 1976.

24. See Karen Orren, *Corporate and Social Change: The Politics of the Life Insurance Industry* (Baltimore: Johns Hopkins Press, 1974); and Serber, "Regulating Reform."

25. Richard E. Grinnold, "A Follow-Up Study of Permanent Disability under Wisconsin Workers' Compensation" (Ph.D. diss.; Madison: University of Wisconsin, 1976), pp. 24, 148–152.

26. "State Workers' Compensation Laws Compared with Essential Recommendations of the National Commission on State Workers' Compensation Laws as of January 1, 1976," (New York: American Insurance Association).

27. On all of this, see Ginnold, "Permanent Disability," pp. 249–53; and David B. Kalis, "Workmen's Comp: Will Uncle Sam Deal Himself In?" *Occupational Hazards* (December 1975): 35–38; "Summary and Analysis of the Key Provisions of S. 2018: National Workers' Compensation Act of 1975," prepared by Bruce Poyer, Labor Occupational Health Program, Institute of Industrial Relations, University of California, Berkeley, 1975; and S. W. Adams and A. J. Jaffe, "Too Little and Too Late," *Supplemental Studies for the National Commission on State Workmen's Compensation Laws*, vol. 2, pp. 13–66 (hereafter *Supplemental Studies*).

28. Alfred M. Skolnik and Daniel N. Price, "Workmen's Compensation Payments and Cost, 1972," *Social Security Bulletin* (January 1974): 30–34.

29. Adams and Jaffe, "Too Little and Too Late."

30. *Compendium on Workmen's Compensation*, pp. 165, 175; and Ginnold, "Permanent Disability," chapter 8.

31. Mary Quartiano, letter to author, August 8, 1977.
32. See William F. Kinder, Director, McKinsey & Co., Los Angeles, "A Look at the Leaders: Time for the Turnaround?" *Best's Review* (July 1976): 78; and compare with *Best's Aggregates and Averages (Property-Liability)*, 1975 ed., p. 1. Also "Sudden Riches for the Casualty Insurers," *Business Week* (May 1, 1978): 66–71.
33. Price, "Compensation Costs, 1974."
34. E. F. Cheit, *Injury and Recovery in the Course of Injury* (New York: Wiley, 1963), p. 264.
35. Serber, "Regulating Reform."
36. *Analysis of Workers' Compensation Laws, 1976*, pp. 26, 27.
37. *Litigation in Workers' Compensation—A Report to the Industry* (San Francisco: California Workers' Compensation Institute, 1974); and Ginnold, "Study of Permanent Disability," pp. 268–75.
38. *Evaluating Disability under the Wisconsin Workmen's Compensation Act* (Employers' Mutual of Wausau, 1971), as cited in Ginnold, "Study of Permanent Disability," p. 195.
39. Ginnold, "Study of Permanent Disability," pp. 199, 200.
40. Mary Quartiano, letter to author, November 21, 1975.
41. Robert A. Fowler, interview with author, April 1975.
42. Ginnold, "Study of Permanent Disability," p. 159.
43. Ibid., p. 194.
44. Benjamin Marcus, "Advocating the Rights of the Injured," in E. F. Cheit and M. S. Gordon, eds., *Occupational Disability and Public Policy* (New York: Wiley, 1963), p. 79.
45. Melville Dalton, "The Role of Supervision," in A. Kornhauser, *Industrial Conflict* (New York: McGraw-Hill, 1954).
46. Louise B. Russell, "Compromise and Release Settlements," *Supplemental Studies*, vol. 3, pp. 188, 195.
47. *Current Developments Section, Daily Labor Report*, Bureau of National Affairs (December 28, 1973): 9–14, "Report on the National Workers' Compensation Standards Act, with Reactions from Insurance Carriers and Corporate Sources"; and "Employers Organize to Fight Federal Standards for Workers' Compensation," *Current Developments Section, Daily Labor Report*, Bureau of National Affairs (February 2, 1974): 1–3.
48. Ralph R. Adams, director of unemployment and workers' compensation, General Motors, and chairman of the Task Force on Workmen's Compensation of the National Association of Manufacturers, (NAM), speech to NAM, conference on workmen's compensation legislation, full text in *Daily Labor Report*, Bureau of National Affairs, February 20, 1974. This estimate is close to my estimate of the true costs of occupational disability to workers.
49. Monroe Berkowitz, "Sources of Information about Workmen's Compensation Recipients," *Supplemental Studies*, vol. 2, pp. 119, 120.

50. Ashford, *Crisis in the Workplace*, chapter 12; Stephen H. Wodka, "Pesticides since Silent Spring," in Garrett de Bell, ed., *The Environmental Handbook* (New York: Ballantine, 1970), pp. 76–79; Ephraim Kahn, "Pesticide Related Illness in California Farm Workers," *Journal of Occupational Medicine* (October 1976), pp. 693–96.

51. See, for example, Lorin E. Kerr, "Coal Workers and Pneumoconiosis," *Archives of Environmental Health* (April 1968), reprinted in 1968 Senate hearings, pp. 797, 798; Willard S. Randall and Stephan D. Solomon, *Building 6, The Tragedy at Bridesburg* (Boston: Little, Brown, and Company, 1977); and Paul Brodeur, *Expendable Americans* (New York: Viking Press, 1974).

52. "Simple Test May Establish Carcinogenicity," *Chemical and Engineering News* (December 22, 1975): 19, 20.

53. For information about this law and the struggle to pass it, see Joel Swartz, "The Fight for Legal Protection," *World Magazine* (July 31, 1976): 3, 6; N. Buder and L. Billings, "Legislation to Control Toxic Substances," *Sierra Club Bulletin* (November–December 1975): 26–31; and "President Signs Law to Control Toxic Substances," *San Francisco Chronicle*, October 13, 1976, p. 7.

54. See Andrew Kalmykow, Assistant General Counsel, American Insurance Association, 1970 Senate hearings, p. 349, for the first figure; the second was told to me in a telephone conversation, August 12, 1976.

55. Lee Holmes, Counsel, American Mutual Insurance Alliance, 1970 Senate hearings, pp. 359–73 represents the first, while *Safety Subjects* (Washington, D.C.: U.S. Department of Labor, Bureau of Labor Standards, Bulletin no. 67, revised 1956), p. 33 argues the second.

56. "Pure Premium Review Sheet, 1977," proposed rates for 1977, (San Francisco: California Compensation Rating Bureau, 1976).

57. G. B. Lemke, Vice-President, Safety and Health Services, Employers' Mutual of Wausau, Wausau, Wisconsin "Loss Cost Analysis and Control," speech delivered at the American Society of Safety Engineers conference, Tampa, Florida, June 28, 1971; mimeographed.

58. Andrew Kalmykow, letter to Robert A. Harris, Counsel, Subcommittee on Labor, U.S. Senate Committee on Labor and Public Welfare, July 11, 1968, reprinted in 1968 Senate hearings, pp. 812–15.

59. See Wirtz, 1969 Senate hearings, pp. 134–36; and *Compendium on Workmen's Compensation*, pp. 284, 285, 289, 290.

60. K. S. Rosten, Manager of Technical Programs, American Pulpwood Association, 1972 House Oversight hearings, p. 389.

61. Lemke, "Loss Cost Analysis and Control," p. 8.

62. John F. Burton, interview with author at School of Business, University of Chicago, Chicago, Illinois, March 13, 1973.

63. *Report of the National Commission*, letter of transmittal, signed by John F. Burton, July 31, 1972.
64. Andrew Kalmykow, American Insurance Association, "Testimony on S. 2018, The National Workers' Compensation Act of 1975," February 2, 1976, available from the American Insurance Association. See also *Firemen's Fund American Companies 1975 Annual Report*, p. 14.
65. Stuart Eisenstat, Charles L. Schultze, and Bert Lance, "Memorandum to the President," May 27, 1977, reprinted in *UAW Washington Report* (July 18, 1977): 2.
66. "Illinois Workers' Comp Law—What Next?" *CACOSH Health and Safety News* (Chicago, July 1977): 2; and *Cumulative Injury—The Problem Can Be Solved* (Industrial Indemnity—A Crum and Forster Insurance Company San Francisco, 1977).
67. George Taylor, staff economist, Standing Committee on Health and Safety, AFL-CIO, Washington, D.C., telephone conversation with author, July 25, 1977.
68. For a history of these efforts see *CACOSH Health and Safety News* (Chicago), May 1976, July 1977; "Brown Lung Unit Members Protest," *Greenville News* (Greenville, S.C.), April 28, 1977; and "Federalizing Workmen's Comp," *Business Week* (February 16, 1974): 85. See also Douglas A. Fraser, President, International Union, United Auto Workers, letter to Secretary of Labor Ray Marshall.
69. Mary Jane Fisher, "Carter Sees 'Operational' Role for Insurers under NHI Plan," *The National Underwriter* (Property and Casualty Insurance Edition), July 16, 1976, p. 1; Edmund Faltenmeyer, "Ever Increasing Affluence Is Less of a Sure Thing," *Fortune* (April 1975), pp. 62 ff; and Charles L. Schultze, Chairman, Council of Economic Advisers, "The Public Use of Private Interest," *Harper's* (May 29, 1977): 43–62.

Chapter 4

1. For information on the IHF see Donald Whorton, *Byssinosis* (Occupational Health Project, Medical Committee for Human Rights, 1972); *Accident Prevention Manual for Industrial Operations*, 6th ed. (Chicago: National Safety Council, 1969), pp. 316, 317; George Perkel, Research Director, Textile Workers Union of America, testimony in 1970 Senate hearings, p. 600; Steve Cohen, memo to Steve Harrison, "Re Industrial Health Foundation," August 20, 1971 (both worked for one of the Nader groups in Washington, D.C.); Industrial Health Foundation,

"Announcing Our New Name But We Are Still IHF" (20-page pamphlet published in Pittsburgh, Pennsylvania, 1970); Rachel Scott, *Muscle and Blood* (New York: E. P. Dutton & Co., 1974), pp. 174–203. For more information on the AOMA see Duane L. Block, president, Industrial Medical Association (now American Occupational Medical Association), 1970 Senate hearings, pp. 767–83.

2. See testimony of George L. Gorbell, Manager of Personnel Safety, Monsanto Corporation and representative of the Manufacturing Chemists Association, 1970 Senate hearings, p. 610; also testimony of D. V. Dernehl, physician, Union Carbide Corporation; and John C. Logan, Universal Oil Products Company and chairman, Manufacturing Chemists Association, 1968 House hearings, p. 349.

3. On the concept of an operational code see Harold L. Wilensky, *Organizational Intelligence* (New York: Basic Books, 1967), pp. 22, 23.

4. On the importance of ideology in controlling a society see Carl Boggs, *Gramsci's Marxism* (London: Pluto Press, 1976), especially the chapter "Ideological Hegemony and Class Struggle."

5. See Francis L. LaQue, *Annual Report of the President of the American National Standards Institute*, 51st Annual Meeting, reprinted in 1970 Senate hearings, p. 480.

6. National Association of Manufacturers, 1970 Senate testimony, p. 1127.

7. *Accident Prevention Manual, 1969*, pp. 7, 8.

8. Ibid., pp. 8–10.

9. Ibid., pp. 5, 6.

10. Ibid., pp. 7, 8.

11. Ibid., pp. 8–10.

12. See George Alvin Cowee, *Practical Safety Methods and Devices* (New York: Van Nostrand, 1916), p. 3.

13. See Ralph Nader, 1968 Senate testimony, p. 587.

14. *Report to the Nation, 1976* (Chicago: National Safety Council), p. 28.

15. See Ralph Nader, *Unsafe at Any Speed* (New York: Grossman, 1965), p. 7.

16. See Donald M. Johnson, chairman, American Insurance Association, *Report of the Chairman, 1974*, p. 62; and Andrew Kalmykow, "Statement of American Insurance Association on S.2018, 'The National Workers' Compensation Act of 1975,' " February 2, 1975, mimeo; and letter from Andrew Kalmykow to author, December 10, 1976. The AIA does not publish an annual report.

17. "Insurance Premium Distribution," *Best's Review*, July 1976, p. 10; and *Sixtieth Annual Report, 1975* (San Francisco: California Inspection Rating Bureau), p. 18.

18. See Donald L. Peyton, 1970 Senate hearings, pp. 458, 459.
19. Nicholas A. Ashford, *Crisis in the Workplace* (Cambridge, Mass.: MIT Press, 1976), pp. 240, 244.
20. LaQue, *ANSI Annual Report*, p. 479.
21. As reprinted in 1970 Senate hearings, pp. 1176, 1177.
22. Donald L. Peyton, letter to Senator Harrison A. Williams, Jr., December 3, 1969, in 1970 Senate hearings, pp. 1206, 1207.
23. See Monica Hollstein, "The Concept of Threshold Limit Values— Scientific and Philosophical Considerations," photocopy, School of Public Health, University of California, Berkeley, December 17, 1972; also E. J. Baier, "Standard for Exposure to Airborne Contaminants," *Archives of Environmental Health*, December 1965, as quoted in Hollstein, p. 2; also Alexander V. Roschin and L. A. Tinofeevskaya, "Chemical Substances in the Work Environment: Some Comparative Aspects of USSR and US Hygienic Standards," *Ambio*, Swedish Royal Academy of Sciences, 1975.
24. Figure arrived at by number of chemicals listed in *Chem Sources* (Flemington, N.J.: Directories Publishing Company, 1958 and 1975 editions). *Chem Sources* bills itself as "the most complete chemical sources guide published on the organic and inorganic products of the U.S. chemical industry."
25. V. N. Fomenko, "Long-Term Effects of Exposure to Toxic Substances," from *Methods Used in the USSR for Establishing Safe Levels of Toxic Substances* (Geneva: World Health Organization, 1975), pp. 75-85.
26. *Monitor*, Labor Occupational Health Program, Berkeley, California, August–September and November–December 1976 issues; for a similar case see "The 'Phosvel Zombies,' " *Newsweek*, December 13, 1976, p. 38.
27. For information on how to negotiate the proceedings, see David Mallino, *A Policy Handbook of Occuaptional Safety and Health Standards* (Washington, D.C.: Government Research Corporation, January 1973).
28. See Irving J. Selikoff, 1970 Senate testimony, pp. 1072-85, and Donald Hunter's *Diseases of Occupations*, 4th ed. (Boston: Little, Brown and Company, 1969), p. 972.
29. From Montague Murray, *Departmental Committee Hearing, 1906*, as pointed out by Irving J. Selikoff in letter to author, June 7, 1976.
30. This history is contained in David Kotelchuk, "Your Job or Your Life," *Health-Pac Bulletin*, March 1973.
31. Ibid.
32. Selikoff, letter to author, June 7, 1976.
33. See Paul Brodeur, *Expendable Americans* (New York: Viking, 1974).
34. See David Kotelchuk, "Asbestos—Science for Sale," *Science for the People*, September 1975, pp. 9-11.

35. Ibid.
36. The following account of the Tyler, Texas, struggle is from Brodeur's *Expendable Americans*.
37. Ibid.
38. Gershon Fishbein, *Occupational Health and Safety Letter*, December 22, 1976, p. 2.
39. See Barry Castleman, "The Export of Hazardous Factories," available for $10.00 in photocopy from Castleman at 1738 Riggs Place, N.W., Washington, D.C. 20009.
40. Kotelchuk, "Your Job or Your Life."
41. Quotation from Paul Brodeur, "No Tangible Effect on Sales and Earnings," *The New Yorker*, November 19, 1973.
42. See, for example, Stan Tulledo and George Thurlow, articles on asbestos in the *Lompoc Record*, Lompoc, California, December 3, 1975; March 3, 1976; March 9, 1976; March 29, 1976; and April 2, 1976; "Workers at Asbestos Plant Sue Coast Company and Its Doctor," *New York Times*, April 18, 1976; *Labor Pulse* (Northern California's Independent Labor Newspaper), March 1976; "Rally and Rites for Asbestos Victims," *San Francisco Chronicle*, June 30, 1976, p. 11.
43. As quoted in article by Robert E. Bedingfeld, "Coordinator of Psychology and Asbestos," *New York Times*, Financial Section, July 15, 1973.
44. Janet B. Schoenberg and Charles A. Mitchell, "Implementation of the Federal Asbestos Standard in Connecticut," *Journal of Occupational Medicine*, December 1974, pp. 781–84; and *Asbestos: The Hazards of Sprayed Fireproofing*, available for 25 cents from Albert DeCosse, c/o Asbestos Workers Local #34, 708 South 10th Street, Minneapolis, Minnesota 55404. See also Robert A. Fowler and Phillip L. Polakoff, *Plain Talk About Asbestos*, and *Asbestos Dust: Everyone's Problem*, available for $1.00 each from WIDES, 2001 Dwight Way, Berkeley, California 94704.
45. Barry Castleman, letter to author, October 2, 1977; Gershon Fishbein, *Occupational Health and Safety Letter*, December 22, 1977, p. 3; and Castleman, "Comment," *Product Liability Insurance*, hearings before the Subcommittee for Consumers, Committee on Commerce, Science, and Transportation, U.S. Senate, April 27-29, 1977, pp. 418-22.
46. This and the next two extracts are taken from Harriet L. Hardy, "Beryllium Poisoning—Lessons in Control of Man-Made Disease," *New England Journal of Medicine*, November 25, 1965, pp. 1188-1200.
47. Herbert K. Abrams, "Diatomaceous Earth Pneumoconiosis—Some Sociomedical Observations," *American Journal of Public Health*, May 1954, pp. 592-99; and Tulledo and Thurlow, *Lompoc Record* articles.

48. *Transactions of the National Conference on Cotton Dust and Health*, May 2, 1970, pp. 89-91, as quoted in Whorton, *Byssinosis*.

49. Quotes on PCBs are from Barry Commoner, director, Center for the Biology of Natural Systems, Washington University, St. Louis; "Labor's Stake in the Environment/The Environment's Stake in Labor," speech, November 28, 1972, Berkeley, California, published in the pamphlet *Jobs and the Environment: Whose Jobs? Whose Environment?* (Berkeley: University of California, Insttute of Industrial Relations, 1973); Vos, *Environmental Health Perspectives*, April 1972, p. 105, as quoted in Commoner, "Labor's Stake in the Environment," pp. 11, 12.

50. See Alice Hamilton, *Exploring the Dangerous Trades* (Boston: Little, Brown and Company, 1943).

51. Joseph A. Page and Mary-Win O'Brien, *Bitter Wages* (New York: Grossman, 1973), pp. 88-94, 197-99.

52. Harriet L. Hardy, "Risk and Responsibility—A Physician's Viewpoint," *New England Journal of Medicine*, October 16, 1975.

53. Quote is from Leon Lewis, "Medical Care under Workmen's Compensation," in *Occupational Disability and Public Policy*, Earl F. Cheit and Margaret S. Gordon, eds. (New York: Wiley, 1963), pp. 124-57.

54. Little research has been done on this. See Dr. I. E. Buff's account of the suppression of research on cotton dust disease in North Carolina in the 1970 Senate hearings, pp. 581, 582.

55. See Daniel M. Berman, Arthur Button, and Mike Ryan, *Crane Corporation, Health and Safety Hazards*, mimeo, 1971, which contains the results of the poll.

56. See Irving R. Tabershaw, "How Is the Acceptability of Risks to the Health of the Workers to Be Determined?" *Journal of Occupational Medicine*, October 1976, pp. 674-76.

57. In order to secure employment, applicants with disqualifying conditions have to lie and are subject to immediate dismissal if found out. In addition, of course, they run the risk of reinjuring themselves.

58. Clarence D. Selby, "Studies of the Medical and Surgical Care of Industrial Workers," U.S. Public Service Bulletin No. 99, as quoted in Henry B. Selleck, *Occupational Health in America* (Detroit: Wayne State University Press, 1962), p. 106.

59. Ibid., p. 103.

60. Tabershaw, "Acceptability of Risks," p. 675.

61. Shepard, *The Physician in Industry*, p. 1.

62. See Miles Colwell, vice-president, Health and Environment, Aluminum Company of America, "Occupational Safety and Health—The Role of Voluntarism," *Journal of Occupational Medicine*, February 1974, pp. 102-06; Robert E. Eckhardt, direc-

tor, Medical Research Division, Esso Research and Engineering Company, "Annals of Industry—Noncasualties of the Work Place," *Journal of Occupational Medicine*, July 1974, pp. 472–77; also Clifford H. Keene, "The Credibility of Occupational Medicine," *Journal of Occupational Medicine*, May 1974, pp. 309–12.

Organized industrial medicine has reacted to public criticism by claiming that documented episodes of malfeasance, such as the Tyler, Texas, asbestos scandal, are unfortunate exceptions and that charges of the existence of a "medical-industrial complex" are baseless. Although none have disputed the basic facts of the Tyler episode, no industrial physicians or medical societies have publicly condemned the Pittsburgh-Corning physicians involved.

63. See Cover, *Journal of Occupational Medicine*, August 1976; also *The Union Voice*, UAW Local 6, March 1 and 8 and August 23, 1974; published in Stone Park, Illinois.

64. *Status Report—Educational Opportunities in Safety*, American Society of Safety Engineers, July 1969, reprinted in 1969 House hearings, pp. 526–70.

65. See George Clayton, Executive Secretary, American Industrial Hygiene Association, 1968 Senate hearings, pp. 477–82.

66. *Membership List, 1970–71*, ACGIH, p. 90.

67. See George Clayton, American Industrial Hygiene Association, 1970 Senate hearings, pp. 165–81; E. J. Baier, part chairman, ACGIH, letter to Senator Harrison A. Williams, September 30, 1969, as reprinted in *1970 ACGIH Transactions*, pp. 245–47.

68. *1970 ACGIH Transactions*, pp. 48, 49.

69. Ibid., p. 46.

70. William Steffan, Occupational Health Section, California State Department of Health, telephone interview with author, October 23, 1973; also Sally Koris, "Industrial Hygienists in Strong Demand as U.S. Widens Role in Workplace Safety," *Wall Street Journal*, August 2, 1977, p. 8.

71. On code of ethics in AIHA see: James L. Weeks, letter to William T. Keene, August 30, 1976 and letter to Charles H. Powell, December 15, 1976 (and repeated in conversation with author, January 9, 1978, Boston, Massachusetts); Charles H. Powell, "Law Committee Report on the Code of Ethics made to AIHA Board of Directors," February 18, 1977; Eileen Crenshaw, "Ethics in the Practice of Industrial Hygiene," AIHA Tri-Sectional Meeting (Delaware Valley, New York, New Jersey; December 3, 1976), and conversation with author, December 18, 1977, Philadelphia.

72. See John Taylor, "The Health Empire," February 13, 1973, in *Control, Conflict, and Change*, no. 3, Detroit, pp. 7–9.

73. *Occupational Health Survey of the Chicago Metropolitan Area* (Cincinnati, Ohio: U.S. Public Health Service, Bureau of Occupational Safety and Health, 1969), pp. 32, 51.

74. In 1970 Senate hearings, p. 178.
75. See "What Occupational Health Problems Confront Us Today?" in *1970 ACGIH Transactions*, p. 100.
76. Ashford, *Crisis in the Workplace*, pp. 442, 443; see also *Medical Manual*, Standard Oil (Indiana).
77. See Marjorie J. Keller, Associate Professor, Department of Community Health Nursing, Virginia Commonwealth University, March 1, 1973, letter to author.
78. See A. C. Blackman, 1970 Senate testimony, p. 777.
79. Ibid., p. 778.
80. See Hugh M. Douglas, "Management of Hazards—The Key to Resources Conservation," *Occupational Hazards*, February 1973, pp. 28-30.
81. Recruiting pamphlet, National Safety Management Society, P.O. Box 14092, Benjamin Franklin Station, Washington, D.C. 20044.
82. See Charles L. Gilmore, Superintendent of Loss Prevention, Monsanto, Texas City, Texas, *Accident Prevention and Loss Control* (New York: American Management Association, 1970), pp. 129, 130.
83. Taylor, "The Health Empire," pp. 8, 9.
84. "OSHA Inspection of Your Plant: Help or Hindrance?" *Occupational Hazards*, November 1972, pp. 27, 28.
85. "Viewpoint: Use OSHA to Save Money, Sell Safety," *Occupational Hazards*, November 1972, pp. 59, 60.
86. On the politics of safety engineering see B. Gwain Bonner, "A Pro's Analysis," *Trial*, July-August 1973, and "Certified Safety Professional—What It Takes to Become One," *Occupational Hazards*, July 1973, pp. 57-59.
87. For general sources on the precise meaning of professionalization see Eliot Friedson, *Profession of Medicine* (New York: Dodd, Mead and Company, 1970), chapter 16.
88. See also Corinne L. Gilb, *Hidden Hierarchies: The Professions and Government* (New York: Harper & Row, 1966).
89. *Medical Manual*, April 1972 (revised), Standard Oil Company (Indiana), including "Approved Organization—1973, Medical and Health Services."
90. See Ray Davidson, *Peril on the Job* (Washington, D.C.: Public Affairs Press, 1970), pp. 54, 104, 122; "Hazards in the Industrial Environment," a conference sponsored by the District 4 Council, OCAW, Houston, Texas, February 20, 21, 1970, pp. 21, 22, 34-36, 53, 54. See also "Hazards in the Industrial Environment," a conference sponsored by District 7 Council, OCAW, Fort Wayne, Indiana, October 24-26, 1969, pp. 20-28.
91. "Hazards in the Industrial Environment," Houston conference, p. 21.
92. Ibid., pp. 53, 54.
93. *Medical Manual*, pp. 21-23.

94. Copy of memo "From the Office of Medical Director, Industrial Hygiene, signed by Paul D. Halley, a report on visit to Amoco Chemical, Joliet plant, on August 16, 1962 by Mr. P. D. Halley and Mr. A. Salazar."

95. See William Curran, "Clinical Information and Records—Public or Confidential?" *New England Journal of Medicine*, June 4, 1964, pp. 1240, 1241.

96. See James M. McNerney, Staff Toxicologist, American Petroleum Institute, to members of the API Committee on Medicine and Environmental Health, July 22, 1971; and memo from P. D. Halley to Dr. Peter M. Wolkonsky, August 19, 1971, "Re—the API Conf. on Heat Stress."

97. W. J. Graham, Division Manager, Employee Relations, The Budd Company, Trailer Division, memo: "Safety Policy for Occupational Safety and Health Act (OSHA) Inspections," effective May 1, 1973, reissued March 1, 1977.

98. Jeanne Stellman and Susan Daum, *Work Is Dangerous to Your Health* (New York: Vintage, 1973).

99. See Willard S. Randall and Stephan D. Solomon, *Building 6: The Tragedy at Bridesburg* (Boston: Little, Brown and Company, 1977).

100. On the Environmental Sciences Laboratory program see: "Science Center's 1973 Targets in Occupational Health," *Occupational Hazards*, December 1972, pp. 39–42; interview with Dr. Irving J. Selikoff; also *IUD Spotlight on Safety and Health*, 3rd quarter 1973, p. 3, by Industrial Union Department, AFL-CIO, Washington, D.C.

101. See "The Society of Occupational and Environmental Health," *Journal of Occupational Medicine* 15: 3 (March 1973).

102. Ibid.

103. On the Shipyard Health Conference see *Occupational Safety and Health Reporter*, "Current Report," January 3, 1974, p. 973, confirmed by a conversation with a person who attended the conference; and Society for Occupational and Environmental Health, "Supplement to Program and Conference Preprints," *International Shipyard Health Conference*.

104. Available for $2.00 from New York Academy of Sciences, 2 East 63rd Street, New York, N.Y. 10021.

105. On the political direction of SOEH see: Society for Occupational and Environmental Health, "Workshop on Directions for Occupational Health in the United States," at the New York Academy of Sciences, November 2, 1973; and "Society for Occupational and Environmental Health, Council Meeting Minutes," meeting occurring October 2, 1973.

Chapter 5

1. In a "closed shop" a worker must join a union to be hired, if one exists. "Right-to-work" laws forbid the closed shop. The term *right to work* was invented by management publicists. On recent labor history see R. O. Boyer and H. W. Morais, *Labor's Untold Story* (New York: Cameron Press, 1955); James J. Matles and James Higgins, *Them and Us* (Englewood Cliffs, N.J.: Prentice-Hall, 1974); and Sidney Lens, *The Labor Wars* (Garden City, N.Y.: Anchor Books/Doubleday, 1973).

2. "AFL-CIO Will Press Legislative Drive to Assist It in Unionizing More Workers," *Wall Street Journal*, February 23, 1977, p. 9.

3. See, for example, John D. Williams, "Big Shippers Push for Decontrol of Truck Industry," *Wall Street Journal*, July 6, 1977, p. 14.

4. "On a Raid with U.S. Agents—Nabbing 29 'Illegals' in One Illinois Town," *U.S. News and World Report*, July 4, 1977, pp. 33, 34; Jane Kramer, "Profiles (Henry Blanton—Part II)," *The New Yorker*, June 6, 1977, p. 66; Sidney Lens, "U.S. Unions: Defeating Themselves," *San Francisco Sunday Examiner*, Sec. A., April 17, 1977, p. 18 (Pacific News Service Release); and "The Next Big Leap in Electronics," *Business Week*, October 24, 1977, pp. 94C-94L.

5. Interviews with Morris Davis, Director, Labor Occupational Health Program, Berkeley, California, autumn 1977.

6. Neal Q. Herrick and Robert F. Quinn, "The Working Conditions Survey As a Source of Social Indicators," *Monthly Labor Review*, April 1971, p. 9.

7. *Analysis of Workers' Compensation Laws* (Washington, D.C.: U.S. Chamber of Commerce, 1976), pp. 24, 25.

8. Max S. Wortman and Wilson Randle, *Collective Bargaining: Principles and Practices*, 2nd ed. (Boston: Houghton Mifflin Company, 1966); and Daniel M. Berman, "Occupational Diseases and Public Policy: The Case of Industrial Lead Poisoning in Missouri," typescript, May 1970. Also based on telephone conversation with John J. Sheehan, legislative director, United Steelworkers of America, Washington, D.C., December 10, 1975, and letter to author, August 6, 1976.

9. Paul Gliddens, *Standard Oil Company (Indiana): Oil Pioneers of the Middle West* (New York: Appleton-Century-Crofts, 1955), p. 28, as quoted in Richard Engler, "Oil Refineries," *Health/Pac Bulletin*, November–December 1974, pp. 7-20.

10. Richard Engler, *Oil Refinery Health and Safety Hazards (Their Causes and the Struggle to End Them)*, available for $3.00 from PhilaPOSH, Room 607, 1321 Arch Street, Philadelphia, Pennsylvania 19107, 1975, p. 1.

11. Harvey O'Connor, *History of the Oil Workers Union-CIO* (Den-

ver: Oil, Chemical, and Atomic Workers Union, 1950), as quoted in Engler, "Oil Refineries," and A. F. Grospirin, "The Control of Chemical Hazards in the Petroleum Industry," in *The New Multinational Health Hazards*, ed. Charles Levinson (Geneva: International Chemical Workers Federation, 1975), pp. 244–50.

12. Ray Davidson, *Peril on the Job* (Washington, D.C.: Public Affairs Press, 1970).

13. Later published as *Expendable Americans* (New York: Viking, 1973).

14. Anthony Mazzocchi, now OCAW's vice-president, made that point in a talk at the University of California, Berkeley, March 15, 1977. See also "Three Shell Oil Plants on West Coast Struck by Oil Workers Union," *Wall Street Journal*, January 26, 1973; and Engler, *Oil Refinery Health*, p. 35.

15. See Howard Kohn, "Malignant Giant: The Nuclear Industry's Terrible Power and How It Silenced Karen Silkwood," *Rolling Stone*, March 27, 1975; and "The Karen Silkwood Case: Death, Plutonium Smuggling and the Curtain of Silence," *Rolling Stone*, January 13, 1977.

16. According to Morris Davis, Director, Labor Occupational Program, Berkeley, California, conversation with author, February 15, 1977; and *Lifelines* (OCAW Health and Safety News), November 1973.

17. This section based on talks with various rank-and-filers and UAW staffers.

18. See then president Leonard Woodcock's speech to Special Collective Bargaining Convention, Detroit, March 22, 1973, mimeo.

19. See Jo Thomas, "Safety Issue to Loom Large in Auto Talks," *Detroit Free Press*, June 24, 1973.

20. See James Harper and Jo Thomas, "Two Angry Workers Shut Chrysler Plant," *Detroit Free Press*, July 25, 1973.

21. See Jo Thomas and Ralph Orr, "New Wildcat Hits Chrysler," *Detroit Free Press*, August 9, 1973.

22. See Ralph Orr, "UAW Finds 'Distressingly Bad' Conditions in Plants," *Detroit Free Press*, August 15, 1973; Jo Thomas and Ralph Orr, "UAW Targets Chrysler—25 Days Left—Job Conditions Emphasized," *Detroit Free Press*, August 22, 1973; also *Collective Bargaining for Occupational Health and Safety* (Berkeley, California: Labor Occupational Health Program, 1973).

23. *UAW Occupational Health and Safety*, November–December 1976, and Dan McLeod, UAW Safety and Health Staff, telephone conversation with author, Detroit, Michigan, March 24, 1977.

24. Comments made in telephone conversation with Ed Jackson, alternate safety representative, UAW Local 560, Milpitas, California, on March 24, 1977; and with Bob Scott, safety representative, UAW Local 1364, Fremont, California, on March 24, 1977.

25. *CACOSH Health and Safety News*, March 1974, July 1974, and October 1976; see also "Auto Workers Local Ends Strike Over Noise at Ford Stamping Plant," *Occupational Safety and Health Reporter*, 1976 (Current Report), pp. 592, 593.
26. *Solidarity*, February 20, 1978; and *Noise Control: A Worker's Manual*, 1978.
27. This account is taken from *Occupational Hazards*, November 1972.
28. *Joint Occupational Health Program Memorandum of Agreement*, United Rubber Workers and B. F. Goodrich Company, 1970; *Joint Occupational Health Program, Joint Memorandum of Agreement*, 1976, United Rubber Workers and xxx Company, and *Article on Environmental Safety and Health*, United Rubber Workers and xxx Company, 1976.
29. See Charles Levinson, Secretary-General, International Chemical Workers Federation, *Work Hazard: Vinyl Chloride* (Geneva: ICF, 1974); also Charles Levinson, ed., *The New Multinational Health Hazards* (Geneva: ICF, 1974), which includes articles by Peter Bommarito, president, URW, "Organized Labor and Industrial Health and Safety," pp. 20–32; and Louis S. Beliezky, "History and Current Developments of the United Rubber, Cork, Linoleum and Plastic Workers of America's Negotiated Joint Occupational Health Program," pp. 92–102.
30. *Teamster Democracy and Financial Responsibility: A PROD Report*, Washington, D.C., May 1976, p. 59; and Warren Morse, Safety and Health Coordinator, Western Conference of Teamsters, Oakland, California, telephone conversation with author, April 1, 1977, and *PROD*, November–December 1974.
31. For example, PROD has given legal support to the Teamsters for a Democratic Union (TDU) and other members who have had elections stolen from them by the leadership, and has sued the international union to recover money donated to aid former president Richard Nixon. See *The Fifth Wheel* (Voice of Northern California Teamsters for a Democratic Union), P.O. Box 23902, Oakland, California 94623; also conversation with Mel Parker, national steering committee member, TDU, Pittsburgh, December 1977.
32. Jean Strickland, "PROD: Caring for Drivers," *Commercial Car Journal*, November 1974, pp. 102–06, and interview with Arthur C. Fox II at PROD office in Washington, D.C., April 19, 1973; telephone conversation with Susan Ginsberg of the PROD staff, April 1, 1977, confirmed by Fox.
33. Daniel M. Berman, *A Job Health and Safety Program on a Limited Budget*, OHP-MCHR, 893 Rhode Island Street, San Francisco, California 94107, 2nd ed., November 1974; and Arthur Button, "Crane Corporation Health and Safety Demands," August 1971, photocopy.

34. See Richard O. Boyer and Herbert M. Morais, *Labor's Untold Story*, p. 45.
35. *Officers' Report to the United Mine Workers of America*, 1973, p. 58.
36. *Officers' Report*, p. 62; Elmer Yocum, 1969 Senate Coal Mine Hearings, pp. 798, 799; and also Harry M. Caudill, *Night Comes to the Cumberlands* (Boston: Little, Brown and Company, 1962).
37. *Officers' Report*, pp. 36, 37, 64, 65.
38. *Fundamentals of Industrial Hygiene*, ed. J. B. Olishifski and F. E. McElroy (Chicago: National Safety Council, 1971), pp. 112, 116.
39. See Lorin E. Kerr, "Coal Workers and Pneumoconiosis," *Archives of Environmental Health*, April 1968, reprinted in 1968 Senate hearings, pp. 797–811; and George Perkel, 1970 Senate hearings, including quotations from the *Industrial Hygiene Digest*, December 1968, p. 600.
40. See *United Mine Workers Journal*, June 1, 1962 and June 15, 1970; and Marilyn K. Hutchinson, "Joint Departmental Administration of Public Law 91-173," *Supplemental Studies for the National Commission on State Workmen's Compensation Laws*, vol. 3, Washington, D.C., pp. 455–63.
41. *United Mine Workers Journal*, November 1, 1972.
42. I. E. Buff, chairman, Committee of Physicians for Miners' Health and Safety; with Dr. Donald L. Rasmussen, chief of Pulmonary Section, Appalachian Regional Hospital, Beckley, West Virginia; and Dr. Hawley A. Wells, Coordinator, Conemaugh Valley Memorial Hospital, Johnstown, Pennsylvania; in 1969 Senate Coal Mine Hearings, pp. 634–68.
43. Bill Worthington, Lee Smith, and Ed Ryan, "Miners for Democracy," in *Rank and File*, ed. Alice and Staughton Lynd (Boston: Beacon Press, 1973), pp. 287–88.
44. By the late sixties many miners had resolved to contest union president Tony Boyle's reign. They were fed up with dictatorial methods, corruption, collaboration with the coal companies, and Boyle's inability to deliver wage increases like his mentor John L. Lewis. The rebels united behind Jock Yablonski's bid for the presidency of the union. Aided by powerful allies such as U.S. Representative Ken Hechler, Ralph Nader, and the black lung doctors, Yablonski made a very strong showing. His supporters believe Boyle stole the election. A few weeks after he lost, Yablonski and his wife and daughter were murdered in their beds. Over the next three years the chain of evidence led inexorably to Tony Boyle, who was ultimately convicted of murder along with several accomplices in the union hierarchy. See Charles McCarry, *Citizen Nader* (New York: Signet, 1973); see also *Officers' Report*, p. 70.
45. Conversation with Bill Worthington, February, 1978.

46. See David Moberg, "Miners Get Third Contract," *In These Times*; and Steven Schneider, "America's Most Dangerous Job," *In These Times*, March 22-28, 1978, p. 3.
47. Falk, "Black Lung Movement."
48. Outside of the coal mines, occupational disease benefits were 1 percent of all compensation benefits, or $47 million in 1974 (see Table 3 in the appendix). Federal black lung benefits for that year totaled $956 million. Daniel N. Price, "Workers' Compensation: Coverage, Payments, and Costs," *Social Security Bulletin*, January 1976, pp. 38-42.
49. Murray V. Hunter, chairman, Joint Industry Health and Safety Committee, 1971 National Bituminous Coal Wage Agreement, arbitration opinion printed in *United Mine Workers Journal*, May 15, 1972.
50. See Irving J. Selikoff, 1970 Senate hearings, p. 1080 for the 1963 figure, and Lorin E. Kerr (UMW occupational health physician) in *United Mine Workers Journal*, September 15, 1972, p. 5 for the 1972 figure.
51. For 1972-1974 these figures are based on unpublished data from the Mine Enforcement Safety Administration.
52. *Officers' Report*, pp. 43, 45.
53. *IAM Guide to Safety and Health Committees*, IAM Safety Department, 1300 Connecticut Avenue, N.W., Washington, D.C. 20036.
54. Thomas F. Mancuso, *Help for the Working Wounded* (Washington, D.C.: IAM, 1976). For a copy send a dollar to *The Machinist*, 909 Machinists Building, Washington, D.C. 20036. Bulk rates are also available.
55. Angelo Cefalo, telephone conversation with author, Washington, D.C., April 13, 1977.
56. *Monitor*, Labor Occupational Health Program, Berkeley, California, February-March 1977.
57. Information assembled from telephone conversations and visits with former safety committee head Robert A. Fowler, A. W. "Fritz" Von Bradford, and a safety committee member at IAM Local 1781, April 1977.
58. Robert A. Fowler, *A Guide for Local Union Health and Safety Committees*, 1974; $2.50 for individuals, $5.00 for organizations; 150 pp. Write to LOHP, 2521 Channing Way, Berkeley, California 94720.
59. "Mike Lefevre (Steelworker)," in Studs Terkel, *Working* (New York: Avon, 1972), pp. 5-10.
60. Rachel Scott, *Muscle and Blood* (New York: E. P. Dutton, 1974), pp. 49-51.
61. This information was gathered in conversations with Katherine Stone, former staffer at the Center for the Study of Responsive Law, Washington, D.C., November 20, 1975; and in a telephone

conversation with John J. Sheehan, December 10, 1975. See also a letter from Sheehan to the author, August 6, 1976.

62. From John J. Sheehan's testimony at the 1972 House Oversight Hearings, p. 616.

63. Boyer and Morais, *Labor's Untold Story*, pp. 102, 103, 312–16; also Nick Migas, "How the International Took Over," in *Rank and File*, ed. Alice and Staughton Lynd (Boston: Beacon Press, 1973), pp. 165–75; and Stanley Aronowitz, *False Promises: The Shaping of the American Working Class* (New York: McGraw-Hill, 1973), pp. 234–38.

64. J. William Lloyd, "Long-Term Mortality Study of Steelworkers, Part V: Respiratory Cancer in Coke Plant Workers," *Journal of Occupational Medicine*, February 1971.

65. Daniel Hannan, telephone conversation with author, Clairton, Pennsylvania, May 1, 1977; and Charles Stokes, telephone conversation with author, Clairton, Pennsylvania, April 30, 1977.

66. Claudia Miller, telephone conversation with author, St. Louis, Missouri, April 26, 1977.

67. *Steel Labor* (official newspaper of the United Steelworkers of America), February 1977, p. 9; Frank Goldsmith, "Coke Ovens, Steel's Big Demon Wins Again," *New Engineer*, January 1977, pp. 25–31; Mary-Win O'Brien, "Dissecting our Demon," letter to *New Engineer*, in reply to Goldsmith article.

68. Steven Wodka, telephone conversation with author, Washington, D.C., May 2, 1977.

69. See letter from J. Bruce Johnstone, for the Coordinating Committee of the Steel Companies, to I. W. Abel, president, United Steelworkers of America, April 9, 1977.

70. See Andrea Hricko, *Working for Your Life: A Woman's Guide to Job Health Standards* (Berkeley, California: LOHP; Washington, D.C.: Health Research Group, 1976), chapter 9. For a copy send $5.00 to LOHP, 2521 Channing Way, Berkeley, California 94720. See also Alice Hamilton and Harriet L. Hardy, *Industrial Toxicology*, 3rd ed. (Acton, Massachusetts: Publishing Sciences Group, 1974), chapter 13.

71. See Hricko, *Working for Your Life*, p. C-6.

72. Coalition for the Medical Rights of Women, 4079A 24th Street, San Francisco, California 94114; "Testimony at OSHA hearings on Lead Level Standards for Pregnant Workers," May 1977, San Francisco; Eva Sullivan, testimony at OSHA lead standard hearings, May 3, 1977. The best article on the legal issues surrounding the exclusion of women from certain work, supposedly because of the health hazards, is Joan T. Samuelson, "Employment Rights of Women in the Toxic Workplace," *California Law Review*, September 1977.

73. See, for example, the Steelworkers' *Safety and Health Manual*, USWA District no. 30, Harry O. Dougherty, director, distributed

at 1974 health and safety conference. Limited number of copies available from Safety Department, USWA, 5 Gateway Center, Pittsburgh, Pennsylvania 15222.

74. Telephone conversation with author, St. Louis, April 26, 1977.

75. *Monitor*, February–March 1977.

76. See articles by Douglas Watson, "Workers in Peril," a four-part series in the *Washington Post*, beginning January 2, 1975.

77. James J. Matles and James Higgins, *Them and Us* (Englewood Cliffs, N.J.: Prentice-Hall, 1974), pp. 12, 153–207.

78. Barry Commoner, *Alliance for Survival* (an address to the 37th International Convention, UE, New York, September 1972), 26 pp.; available for 25 cents from UE, 11 E. 51st Street, New York, N.Y. 10022.

79. "Union and Health Group Fight Lead Poisoning," *UE News*, December 23, 1972, and "Lead Poisoning at Gould, Inc.," Trenton, New Jersey," Report to Local 108, UE, November 14, 1974, Public Citizen's Health Research Group, Room 708, 2000 P Street, N.W., Washington, D.C. 20036.

80. Carolina Brown Lung Association, Columbia Chapter, P.O. Box 545, West Columbia, S.C. 29169.

81. According to J. Gary DiNunno, "J. P. Stevens: Anatomy of An Outlaw," *American Federationist* (Washington, D.C.: AFL-CIO, April 1976), pp. 1–8; confirmed by Eric Frumin, ACTWU Research Staff, telephone conversation with author, Brooklyn, New York, May 16, 1977.

82. Ibid. Also see article by Jeanne Shinto, "The Breathless Cotton Workers," *The Progressive*, August 1977, pp. 27–29.

83. For these figures see *Historical Statistics of the United States, Colonial Times to 1970*, 2 vols. (Washington, D.C.: U.S. Department of Commerce, Bureau of the Census, 1975), series D16, p. 127; and *Statistical Abstract of the United States, 1976* (Washington, D.C.: U.S. Department of Commerce, Bureau of the Census, 1976), p. 372; also Ephraim Kahn, "Pesticide Related Illness in California Farm Workers," *Journal of Occupational Medicine*, October 1976, pp. 693–96; Peter A. Greene, Health Research Group, "A Thanksgiving Disgrace: Farmworker Health and Safety," November 1976, mimeo; and Barry Commoner, *The Closing Circle* (New York: Bantam Books, 1974), pp. 149–51.

84. See Thomas Whiteside, "Tomatoes," *New Yorker*, January 24, 1977, pp. 36ff.

85. Greene, "Thanksgiving Disgrace."

86. "Roberto Acuna, Farmworker," in Studs Terkel, *Working* (New York: Avon, 1972), pp. 30–38.

87. According to Dan Georgakin and Marvin Surkin, *Detroit: I Do Mind Dying* (New York: St. Martin's Press, 1975), p. 80.

88. See Donald L. Ream, Chief, Division of Workmen's Compensa-

230   *Death on the Job*

tion Standards, Employment Standards Administration, U.S. Department of Labor, "Workmen's Compensation Coverage of Agricultural Workers," speech at annual meeting of American Public Health Association, San Francisco, California, November 5-8, 1973; as quoted in Ashford, *Crisis in the Workplace*, p. 522.

89. Lawrence A. Fuchs, *Hawaii Pono: A Social History* (New York: Harcourt, Brace and World, 1961), pp. 206-40; 357-70.

90. See Theodore Rosengarten, *All God's Dangers: The Life of Nate Shaw* (New York: Avon, 1975), for an account of the tribulations of a black farmer in Alabama in the first half of the twentieth century who stood up for his rights.

91. "Roberto Acuna, Farmworker," p. 35.

92. See Harry Bernstein, "Teamsters Withdraw, Leave Field to Chavez," *Los Angeles Times*, March 11, 1977, p. 1.

93. Steven Wodka, "Pesticides Since *Silent Spring*," in *The Environmental Handbook*, ed. Garrett de Bell (New York: Ballantine, 1970), pp. 76-91.

94. The discussion of OSHA, pesticides, and safety in agriculture is based on Ashford, *Crisis in the Workplace*, pp. 141, 529-32; telephone conversation with John Landis, OSHA Region 9 office, San Francisco, May 18, 1977; and Jack Anderson, "Poison in the Fields," *San Francisco Chronicle*, January 23, 1976.

95. See Ephraim Kahn, "Pesticide Related Illness in California Farm Workers," *Journal of Occupational Medicine*, October 1976, pp. 693-96.

96. See Commoner, *Closing Circle*, chapter 9, "The Technological Flaw"; "AB 1192, Social Impact Statements," California State Assembly Bill, 1977-1978 regular session; "AB 1537, Farmworker Mechanization Displacement Fund," pamphlet from United Farmworkers of America, 630 9th Street, Sacramento, California; Richard Levins, "Genetics and Hunger," in *Genetics*, September 1974, pp. 67-76; William Olkowski and Helga Olkowski, "Observations on an Integrated Control Program for Urban Ornamental Plants," copy of talk presented at the 1974 California Golf Course Superintendents' Institute, mimeo; "U.S. Favors Shift from Pesticides" (account of speech by Secretary of Agriculture Bob Bergland), *San Francisco Chronicle*, September 28, 1977, p. 6; and Douglas R. Sease, "Low-Flying Dusters Are Now Highfliers on Nation's Farms," *Wall Street Journal*, June 16, 1977, p. 1.

97. See Mike Cherry, *On High Steel (The Education of an Ironworker)* (New York: Quadrangle, 1974), pp. 76-77.

98. "Hub Dillard, Crane Operator," in Studs Terkel, *Working* (New York: Avon, 1972). pp. 49-54.

99. Cherry, *On High Steel*.

100. Interview with author, Boston, Massachusetts, April 7, 1976.

101. Robert A. Georgine, letter to secretaries, State and Local Building and Construction Trades Councils, AFL-CIO, December 1, 1976, which also contains steps to follow when a hazard is observed.

102. This manual was published jointly by the Building and Construction Trades Department of the AFL-CIO and by the National Contractors association, an industry group.

103. Telephone conversation with Jim Lapping, Health and Safety Coordinator, Building and Construction Trades Department, AFL-CIO, Washington, D.C., June 7, 1977.

104. See, for example, Dr. Samuel S. Epstein, "Corporate Sabotage in the Battle against Industrial Cancer," *In These Times*, April 26–May 2, 1978, pp. 16, 17, from his book *Politics of Cancer*, to be published by the Sierra Club; also "Chemical Dangers in the Workplace," 34th Report, Committee on Government Operations, House of Representatives, September 27, 1976.

105. Richard E. Ginnold, "Worker Attitudes on Occupational Safety and Health," 1976, mimeo, p. 5.

106. Ibid., p. 6.

107. See Stanley Aronowitz, *False Promises*, especially pp. 214–63.

108. See Deborah Sue Yaeger, "Garment Union Tries to Save the Industry (and Also Its Jobs)," *Wall Street Journal*, October 18, 1976, p. 1.

109. William Serrin, *The Company and the Union (The "Civilized Relationship" of the General Motors Corporation and the United Auto Workers)* (New York: Knopf, 1973), quotation from a "labor journalist" as cited in Murray Kempton's review of Serrin's book in *The New York Review of Books*, February 8, 1973, p. 12.

110. Arthur M. Ross, "The Natural History of the Strike," in A. W. Kornhauser, Robert Dubin, and A. M. Ross, eds., *Industrial Conflict* (New York: McGraw-Hill, 1954), pp. 23–26.

111. "Cancer on the Job—'Alarming' U.S. Study," *San Francisco Chronicle*, September 12, 1978, p. 1. This article discusses AFL-CIO plans to establish a health and safety center.

112. See Allen B. Coats, General Representative, Metal Trades Department, AFL-CIO, "Asbestos—The U.S. Navy's Problem," presented at San Francisco Press Club, July 11, 1977, photocopy, p. 4.

113. Dr. Phillip L. Polakoff, "Asbestos-Related Disease Amongst Shipyard Workers" (Mare Island Survey), prepared for Metal Trades Council, AFL-CIO, Mare Island Naval Shipyard, Vallejo, California, July 1977.

114. On all this see, "AFL-CIO Will Press Legislative Drive to Assist It in Unionizing More Workers," *Wall Street Journal*, February 23, 1977, p. 1; Harry Bernstein, "Top Union Leaders Take 'Risk,' Open Campaign for Major Labor Legislation," *Los Angeles Times*, February 23, 1977, Part I, p. 10; and C. Wright Mills, *The Power*

Elite (New York: Oxford University Press, 1956), pp. 264, 265.

115. Telephone interview with John J. Sheehan, Legislative Director, USWA, Washington, D.C., December 10, 1975.

116. See David Jenkins, *Job Power* (Baltimore, Md.: Penguin, 1973), and "Industrial Democracy—It Catches on Faster in Europe Than U.S.," *New York Times*, May 13, 1973, Financial Section, p. 1; Neil Ulman, "The Worker's Voice—Giving Employees a Say in Firms' Management Seen Gaining in Europe," *Wall Street Journal*, February 23, 1973, p. 1; Jack Mundey, "Labor and Environment, The Australian Green Bans," *The CoEvolution Quarterly*, P.O. Box 428, Sausalito, California 94965, Spring 1977, pp. 40–45; and William M. Bulkeley, "Plan to Ban Nonreturnable Bottles, Cans Stirs Ire in Michigan As Vote Approaches," *Wall Street Journal*, October 27, 1976, p. 40.

117. See Irving Bluestone, vice-president, General Motors Division, UAW, "Toward Democracy at the Workplace," National Conference on Social Welfare, Atlantic City, New Jersey, May 29, 1973, mimeo, 17 pp.

Chapter 6

1. See *Historical Statistics of the United States, Colonial Times to 1970*, 2 vols. (Washington, D.C.: U.S. Department of Commerce, Bureau of the Census, 1975); and *Statistical Abstract of the United States, 1976* (Washington, D.C.: U.S. Department of Commerce, Bureau of the Census, 1976).

2. *Historical Statistics*, series D927, 940, pp. 176, 177; *Statistical Abstract, 1976*, p. 384. The percentages were calculated to include agricultural workers in the total employed.

3. Gabriel Kolko, *Wealth and Power in America* (New York: Praeger, 1962).

4. W. W. Rostow, "Caught by Kondratieff," *Wall Street Journal*, March 8, 1977, editorial page; *Business Conditions Digest*, U.S. Department of Commerce, Bureau of Economic Analysis, July 1976, p. 30; "What the Marxists See in the Recession," *Business Week*, June 23, 1975, pp. 86, 87; Lewis Beman, "The Slow Road Back to Full Employment," *Fortune*, June 1975, pp. 84ff. Joseph Kraft, "Labor Facing New Problem," *Boston Globe*, October 7, 1975, p. 23; James C. Hyatt, "Labor's Want List," *Wall Street Journal*, February 14, 1977, p. 1; Andrew Levison, "The Working-Class Majority," *New Yorker*, September 2, 1974; Barry I. Castleman, "The Export of Hazardous Factories," July 1977,

photocopy; available from Barry Castleman, 1738 Riggs Place, N.W., Washington, D.C. 20009; $6.00, 70 pp. To be published.

5. Richard J. Barnet and Ronald E. Muller, *Global Reach* (New York: Simon and Schuster, 1974); Harry Magdoff, *The Age of Imperialism* (New York: Monthly Review Press, 1969); Ronald E. Müller, "The Political Economy of Global Corporations and National Stabilization Policy," paper presented at the American Assembly, Columbia University, New York, December 12-15, 1974; Christopher Lydon, "Jimmy Carter Revealed: He's a Rockefeller Republican," *The Atlantic*, July 1977, pp. 50-57; Jeremiah Novak, "The Trilateral Connection," *The Atlantic*, July 1977, pp. 57-60.

6. Joseph Eyer and Peter Sterling, "Stress-Related Mortality and Social Organization," *The Review of Radical Political Economics*, Spring 1977, pp. 1-44. The *Review* is available from URPE, 41 Union Square West, Room 901, New York, N.Y. 10003.

7. Harry Braverman, *Labor and Monopoly Capital: The Degradation of Work in the Twentieth Century* (New York: Monthly Review Press, 1974), p. 321; Robert L. Simison, "Mass-Output Methods Help Fox and Jacobs Gain Leadership in Housing," *Wall Street Journal*, March 29, 1978, p. 1; Barry Commoner, *The Closing Circle* (New York: Bantam Books, 1974), pp. 168-74; V. N. Fomenko, "Long-Term Effects of Exposure to Toxic Substances," from *Methods Used in the USSR for Establishing Biologically Safe Levels of Toxic Substances* (Geneva: World Health Organization, 1975), pp. 75-85; *Monitor*, Labor Occupational Health Program, Berkeley, California, August-September and November-December 1976 issues.

8. Karl Marx, *Capital* (New York: The Modern Library, 1906 ed.), p. 399; and Morris Davis, "Occupational Hazards and Black Workers," *Urban Health*, August 1977, pp. 16-18.

9. See, for example, Thomas F. Mancuso, "Lung Cancer among Black Migrants," *Journal of Occupational Medicine*, August 1977; Eyer and Sterling, "Stress-Related Mortality"; Asa Christina Laurell, Jose Blanco Gil, Teresa Machetto, Juan Palomo, Claudia Perez Rullfo, Manuel Ruiz de Chavez, Manuel Urbino, and Nora Velazquez, "Disease and Rural Development: A Sociological Analysis of Morbidity in Two Mexican Villages," *International Journal of Health Services*, no. 3, 1977, pp. 401-23.

10. *Work in America (Report of a Special Task Force to the Secretary of Health, Education, and Welfare)* (Cambridge, Massachusetts: MIT Press, 1973), p. xvii.

11. David Jenkins, *Job Power* (New York: Penguin, 1973); Ken Coates, ed., *Can the Workers Run Industry?* (London: Sphere Books Ltd., in association with The Institute for Workers' Control, 40 Park Street, London, W.1, 1968); and E. S. Greenberg, "The Consequences of Worker Participation: A Clarification of the Theoretical

Literature," *Social Science Quarterly*, September 1975, pp. 191–209.

12. Robert A. Fowler, former Carpenters' Union member, interview in San Francisco, July 7, 1976.

13. "Doug Fraser Speaks Out on Jobs, 4-Day Week, Workers' Safety and Health," *UAW Washington Report*, September 26, 1977.

14. *The New Multinational Health Hazards*, ed. Charles Levinson (Geneva: International Chemical Workers Federation, 1975), p. 15.

15. Castleman, "Export of Hazardous Factories," p. 35, quoting from "Overseas Operations of U.S. Industrial Companies, 1976–78," McGraw-Hill Economics Department, 1976.

16. Castleman, "Export of Hazardous Factories," pp. 8–11.

17. *Minerals Yearbook*, U.S. Department of the Interior, Bureau of Mines, articles on asbestos for years 1960 and 1973.

18. Raymundo Arroyo, "Relative and Absolute Pauperization of the Brazilian Proletariat in the Last Decade," in LARU Studies, no. 1, October 1976, pp. 30–36; available from Brazilian Studies, Latin America Research Unit, Box 673, Adelaide St. P.O., Toronto 1, Ontario, Canada.

19. Carlos Afonso, "A Note on the Exploitation of Labour in the 'Brazilian' Automobiles Industries," [sic] and other articles in *Multinationals and Brazil*, ed. Marcos Arruda, Herbet de Souza, and Carlos Afonso (Toronto: Brazilian Studies, Latin America Research Unit, 1975), pp. 30–36.

20. Roberto Suzedelo, "Acidentes de Trabalho—A Queda dos Números," and Tânia Coelho, "O Posto de Reabilitação," both in *Movimento* (Rio de Janeiro), July 11, 1977.

21. Jack Anderson, "The Red Carpet," *San Francisco Chronicle*, August 24, 1977, p. 45.

22. Gerry Pocock, "European Trade Union Conference Sparks Increased East-West Ties," available from Transnational Features Service, P.O. Box 19148A, Los Angeles, California 90019.

23. See Fred Hirsh et al., *An Analysis of Our AFL-CIO Role in Latin America* (or *Under the Covers with the CIA*) (San Jose, California, 1975), and Cynthia Sweeny et al., *CIA & CWA*, published June 1975 by Local 11500 members, c/o P.O. Box 8155, San Diego, California 92102, p. 7.

24. For information on unions in Chile see "Chile: A New Report by the National Union of Miners in Britain," about a trip to Chile between April 30 and May 6, 1977. Available from Transnational Features Service, P.O. Box 19148A, Los Angeles, California 90019; see also "Why U.S. Labor Is Looking Overseas Again," *Business Week*, April 17, 1978, pp. 126–30.

25. See, for example, *Threshold Limit Values in the United States, Germany, and Sweden, 1975* (Geneva: International Chemical Workers Federation, 1975).

26. *The New Multinational Health Hazards*, p. 16.

27. *Chemical Dangers in the Workplace*, 34th Report by the Committee on Government Operations, September 27, 1976 (Washington, D.C.: U.S. Government Printing Office, 1976), pp. 10, 11.

28. "Illinois Workers' Comp Law—What Next?" *CACOSH Health and Safety News*, July 1977; *Cumulative Injury—The Problem Can Be Solved* (San Francisco: Industrial Indemnity—A Crum and Forster Insurance Company, 1977).

29. Bob MacMahon, "Carter Stalls on Cotton Mill Standards," *The Guardian* (New York) June 7, 1978, p. 3; Carol H. Falk, "Job Inspectors Need Warrants, Top Court Rules," *Wall Street Journal*, May 24, 1978, p. 2; and "U.S. Supreme Court Decision in Marshall *v.* Barlow's, Inc., " *Occupational Safety and Health Report* 7, no. 52 (May 25, 1978).

30. Interview with Dr. Phillip L. Polakoff, San Francisco, July 12, 1977; Allen B. Coats, general representative, Metal Trades Department, AFL-CIO, "Asbestos—the U.S. Navy's Problem," photocopied handout prepared July 1977; also *Minerals Yearbook, 1960–1973*, U.S. Bureau of Mines, articles on asbestos, for foreign production; more recently, *Dying for Work—Occupational Health and Asbestos*, pamphlet from North American Committee for Latin America, 207 W. 106th Street, New York, N.Y. 10025; $2.00, esp. p. 15.

31. "The Asbestos Hazard," 1977, pamphlet published by Green Ban Action Committee, 77 School Road, Hall Green, Birmingham 28, Great Britain.

32. *IAM Guide for Safety and Health Committees*, International Association of Machinists, 1300 Connecticut Avenue, N.W., Washington, D.C. 20036.

33. Conversation with members of safety committee of Machinists Local Lodge 1781, Burlingame, California, September 14, 1977. See also "Organizing for Job Safety and Health: A Conversation about COSH Groups," photocopy, February 22, 1978. Send $1.00 to Daniel M. Berman, 893 Rhode Island Street, San Francisco, California 94107.

34. Candace Cohn and Mel Packer, "An Open Letter to PACOSH," summer 1975.

35. Anthony Mazzocchi, testimony as presented by Steve Wodka, *Control of Toxic Substances in the Workplace*, hearings before a subcommittee of the Committee on Government Operations, U.S. House of Representatives, May 11, 12, and 18, 1976, p. 30; and "It's Our Right to Know!" petition circulated by MassCOSH, P.O. Box 17326, Back Bay Station, Boston, Massachusetts 02116; autumn 1977.

36. J. Warren Salmon, "Monopoly Capital and the Reorganization of the Health Sector," *The Review of Radical Political Economics*, Spring 1977, pp. 125–33.

37. Alice Hamilton, *Exploring the Dangerous Trades* (Boston: Little,

Brown and Company, 1943), p. 332; *Institute of Industrial Hygiene and Occupational Diseases* (Moscow, 1974), pamphlet, 56 pages; I. V. Sanockij, "Investigation of New Substances: Permissible Limits and Threshold of Harmful Action," from *Methods Used in the USSR*, pp. 9–18; Andrea Hricko, "Occupational Health in the Soviet Union," presented at American Public Health Association Convention, Session on Occupational Health in Socialist Countries, November 19, 1975; available from LOHP, 2521 Channing Way, Berkeley, California 94720; and George J. Ekel and Warren J. Teichner, *An Analysis and Critique of Behavioral Toxicology in the USSR* (Cincinnati, Ohio: NIOSH, December 1976), 129 pp.

38. "PPG and Soviet Set Pact for Big Plastics Complex," *New York Times*, March 20, 1974.

39. Fomenko, "Long-Term Effects," pp. 77, 78.

40. See John G. Fuller, *The Poison That Fell from the Sky* (New York: Random House, 1978).

41. Morris Davis, telephone conversation with author, October 18, 1977.

42. Rex Cook, president, OCAW Local 1–5, lecture at College of Conservation and Natural Resources, University of California at Berkeley, spring 1976.

43. Inspection data from official inspections can be obtained through the Freedom of Information Act—a good tactic for union organizers.

44. Lorin E. Kerr, occupational health physician, United Mine Workers, "Statement on State Workmen's Compensation to the National Commission on State Workmen's Compensation Laws," September 22, 1971, mimeo.

Appendix 1

Statistical Tables

Table 1

Middle 1960s: Estimated annual expenditures earmarked for prevention
and compensation of occupational casualties, by sector

Category	Private Sector	Public Sector	Total
	million $	*million $*	*million $*
Prevention	1,528[a]	41[b]	1,569
	(34%)[c]	(0.9%)	(35%)
Compensation	2,339[d]	563[e]	2,902
	(52%)	(13%)	(65%)
Total	3,867	604	4,471
	(86%)	(14%)	(grand total)

[a]This figure includes a "guesstimate" of $1,500 million spent by private companies, based on a backward projection of figures in the "First Annual McGraw-Hill Survey of Investment in Employee Safety and Health," May 25, 1973, and an estimated $28.3 million spent by insurance carriers and self-insurers on prevention. This is 1.2 percent of total private insurance premiums, according to *Insurance Expense Exhibit (Countrywide)*, December 31, 1966, compiled by the National Council on Compensation Insurance, New York (available in 1968 Senate hearings, p. 820).

[b]The Department of Labor spent $2,234,000 in the mid-1960s, according to testimony by Willard W. Wirtz, 1968 Senate hearings, p. 204. The Department of Health, Education, and Welfare spent about $6 million, according to David Mallino, "A Policy Handbook on Occupational Safety and Health Standards" Washington, D.C.: Government Research

Center, 1973, photocopy p. 6. According to Jerome Gordon, $23 million was spent on industrial inspection by the states in 1965 (1968 Senate hearings, p. 557). Victoria Trasko estimated that all the official industrial hygiene agencies in the U.S. spent "less than $3,000,000 annually" ("Silicosis: A Continuing Problem," Public Health Reports, September 1958). Of the $570 million employers spent on premiums to state-owned funds and in federal compensation benefits, 1.2 percent is $6.8 million (Alfred M. Skolnik and Julius W. Hobson, Office of Research and Statistics, Social Security Administration, "Workmen's Compensation Payments and Costs, 1965," *Social Security Bulletin,* January 1967, pp. 29–31, 36).

[c]Including $2,100 million in premiums to private insurance carriers and an estimated $255 million spent by self-insurers minus 1.2 percent presumed to be spent on preventive activities, according to Skolnik and Hobson.

[d]Sum of the premiums to state funds and minus 1.2 percent presumed spent on preventive activities.

Table 2

1972: Estimated expenditures earmarked for prevention
and compensation of occupational casualties, by sector

Category	Private Sector	Public Sector	Total
	million $	*million $*	*million $*
Prevention	2,397[a]	110[b]	2,507
	(29%)[c]	(1.3%)	(31%)
Compensation	4,653[d]	1,036[e]	5,689
	(57%)	(13%)	(69%)
Total	7,050	1,146	8,196
	(86%)	(14%)	(grand total)

[a]The "First Annual McGraw-Hill Survey of Investment in Employee Safety and Health" estimated that private employers spent $2,340 million on prevention, to which is added $56.6 million (1.2 percent of the $4.7 billion spent on private insurance and for self-insurance), as noted in Alfred M. Skolnik and D. N. Price, "Workmen's Compensation Payments and Costs, 1972," *Social Security Bulletin,* January 1974, pp. 30–34.

[b]This is a sum of the following figures: OSHA budget, $35.7 million; NIOSH budget, $26.3 million; state programs estimated budgets, $7.8 million in federal money which presumably is matched by state funds,

$23 million spent on state programs in 1965 (assumed to have continued unchanged) (see Nicholas A. Ashford, *Crisis in the Workplace* [Cambridge, Mass.: MIT, 1976], pp. 240–44) and $17.28 million, or 1.2 percent of the cost of state fund and federal compensation.

[c]This consists of the $4.71 billion spent on premiums to private insurance companies for self-insurance, minus $56.6 million presumed to have been spent on prevention (Skolnik and Price, "Workmen's Compensation").

[d]State fund plus federal compensation costs, minus presumed safety expenditures of 1.2 percent, including federal black lung benefits for coalminers (ibid).

Table 3

1974: Estimated expenditures earmarked for prevention
and compensation of occupational casualties, by sector

Category	Private Sector	Public Sector	Total
	million $	*million $*	*million $*
Prevention	3,359[a]	144[b]	3,485
	(32%)[c]	(1.2%)	(33%)
Compensation	5,533[d]	1,423[e]	6,956
	(53%)	(14%)	(67%)
Total	8,892	1,549	10,441
	(85%)	(15%)	(grand total)

[a]"First Annual McGraw-Hill Survey of Investment in Employee Safety and Health," May 25, 1973. Estimated by interpolation of investment planned for 1973 and for 1976 plus $67 million (1.2 percent of the amount spent on private insurance and for self-insurance), from D. N. Price, "Workers' Compensation: Coverage, Payments, and Costs, 1974," *Social Security Bulletin*, January 1976, pp. 38–42.

[b]OSHA budgeted expenditures of $70.4 million for fiscal year 1974; NIOSH spent $31.2 million in 1974; state expenditures: $25.08 million, presumably spent to equal that amount in federal matching funds; $17.28 million (1.2 percent of cost of state fund and federal workers' compensation, as noted in Price, "Workers' Compensation."

[c]Cost of premiums paid to private insurance companies and of self-insurance, minus 1.2 percent presumed spent on prevention (ibid.).

[d]Premiums paid to state compensation funds and costs of federally financed programs for compensation, including black lung compensation for coalminers (ibid.).

Table 4

Workers' compensation and the national economy, 1925-1974

Year	GNP[a]	Employer Costs of Compensation[b]	Compensation Costs as Percent of Covered Payroll[b]	Compensation Costs as Percent of GNP	Prevention plus Compensation Costs as Percent of GNP[c]
	billion $	million $	%	%	%
1925	91.3	250[d]	—	0.27	—
1940	100.6	421	1.19	0.42	—
1946	210.7	726	0.91	0.34	—
1950	284.6	1,013	0.89	0.36	—
1955	397.5	1,532	0.91	0.39	—
1960	503.7	2,055	0.93	0.40	—
1965	684.9	2,908	1.00	0.42	0.66[e]
1970	977.1	4,882	1.13	0.50	—
1972	1,158.0	5,759	1.16	0.50	0.71
1974	1,397.4	7,780	1.24	0.56	0.75

^{a}Gross National Product (GNP) figures are from the U.S. Dept. of Commerce, *Historical Statistics of the United States, Colonial Times to 1970*, and various more recent almanacs.

^{b}*Compendium of Workmen's Compensation* (Washington, D.C.: National Commission on State Workmen's Compensation Laws, 1973), p. 279.

^{c}R. H. Lansburgh, "Forward," *The Annals*, January 1926, p. viii. This figure was borrowed from the text, which stated, "The mere workmen's compensation which is added to the selling price must approach $250,000 annually." This is clearly a very rough estimate.

^{d}See Tables 1-3 for the sources of these estimates.

^{e}This estimate is for the middle 1960s.

Table 5

U.S. workers' compensation benefits, policy year 1972, by type of disability[a, b]

Type of Disability	A No. of Cases	B Compensation Paid	C Av. Compensation Cost (B/A)[c]	D Medical Benefits	E Av. Medical Benefits Cost (D/A)	F Total Cost (B+D)	G Av. Cost, Total Benefits (B+D)/A	H Est. % of Wage Losses Replaced for Cases Entering System[d]	I Est. Uncompensated Wage Loss I = B(100−H)/H
	thousands	million $	$	million $	$	million $	$	$	million $
Medical only	4,261 (80.2%)	0 (0%)	0	191	45	191 (8%)	45	0[e]	341
Temporary total	793 (14.9%)	364 (23%)	460	250	315	614 (25%)	775	44[f]	464
Permanent minor disability	203 (3.8%)	502 (31%)	2,470	217	1,080	721 (29%)	3,552	19[g]	2,140
Permanent major disability	41.2 (0.8%)	518 (32%)	12,600	196	4,760	714 (29%)	17,360	19[g]	2,204
Permanent total disability	1.5 (0.0015%)	56 (4%)	37,200	38	26,600	93 (4%)	63,800	28[h]	143
Death	5.6 (0.01%)	153 (10%)	27,200	7	1,260	160 (6%)	28,460	32[i]	324
Totals	5,305 (100%)	1,593 (100%)	300	900	170	2,493 (100%)	470	22[j]	5,616

242

[a] National Council on Compensation Insurance, "Countrywide Workmen's Compensation Experience Including Certain Competitive State Funds," October 1975. The figures were calculated according to a smaller base than the Skolnik and Price estimates quoted in Table 2.

[b] Excepting Nevada, North Dakota, Ohio, Washington, and Wyoming, all with exclusive state funds.

[c] "Countrywide Workmen's Compensation."

[d] For source of estimates, see Monroe Berkowitz, "Workmen's Compensation Income Benefits: Their Adequacy and Equity," *Supplemental Studies* 1, pp. 189–274.

[e] It is assumed that average loss of work time of two days for each of the 4,261,000 incidents for which medical care only was provided, and two days of work at $40/day were lost; thus the calculation of estimated uncompensated wage loss is $(4,261,000) \times (\$40) \times (2) = \$341,000,000$.

[f] *Compendium on Workmen's Compensation* (Washington, D.C.: National Commission on State Workmen's Compensation Laws, 1973), p. 119, calculated by interpolation of estimates from 54 jurisdictions.

[g] Ibid., p. 139, calculated by interpolation of estimates of income replacement payable to a worker who is 50 percent disabled, for 29 jurisdictions. Unrealistically, for lack of better data, this assumes that the percent of income replacement is the same for those suffering minor and major disabilities.

[h] Ibid., p. 125, calculated by interpolation of estimates from 52 jurisdictions.

[i] Ibid., p. 140, calculated by interpolation of estimates from 51 jurisdictions.

[j] Calculated by summing compensation paid ($1593m) and estimated uncompensated wage loss ($5616m) to arrive at a figure for total wage loss of $7209m. Dividing $1593 by $7209 gives 0.22 or a 22 percent rate of wage replacement for cases entering the workers' compensation system.

Appendix 2

A Short Guide to Worker-Oriented Sources

Since I wrote the first *Guide to Worker-Oriented Sources in Occupational Safety and Health* for the Medical Committee for Human Rights in 1974, there has been a tremendous explosion of readable and technically accurate materials just in English. I could not pretend to list them all and comment intelligently. To find out where to get help, first contact your union if you have one, and then the following sources, asking for their publications lists.

Projects and Publications in the United States

BLACK LUNG ASSOCIATION (BLA)
Bill Worthington, Chairperson
P.O. Box 68
Coxton, Kentucky 40831; telephone: 606-837-3380

BROWN LUNG ASSOCIATION
Brown lung, a disease of cotton textile workers, is epidemic in the Carolinas, but very little is being done by the companies to prevent exposure to cotton dust, and very few

workers have secured compensation. To work on these problems write the Brown Lung Association
  Greenville Chapter
  P.O. Box 334
  Greenville, S.C. 29602; telephone: 803-235-2886
Pamphlet, "Questions and Answers about Compensation for Brown Lung," $.50, includes a list of other Brown Lung Association chapters and publications.

CHICAGO AREA COMMITTEE FOR OCCUPATIONAL SAFETY AND HEALTH (CACOSH)
  542 S. Dearborn, Rm. 508
  Chicago, Ill. 60605; telephone: 312-939-2104
CACOSH *Health and Safety News*, published monthly; subscription $3 per year.

ENVIRONMENTAL SCIENCES LABORATORY
  Mt. Sinai School of Medicine
  100th St. and 5th Ave.
  New York, N.Y. 10029; telephone: 212-650-6500
Most famous center for occupational health research in the United States, under Dr. Irving J. Selikoff's leadership. Call them if you have a big, mysterious problem.

HEALTH RESEARCH GROUP
  Sid Wolfe
  2000 P St., NW, Suite 708
  Washington, D.C. 20036; telephone: 202-872-0320
One of Nader's best groups, capable of sometimes brilliant lobbying and technical aid.

INTERNATIONAL ASSOCIATION OF MACHINISTS
  Health and Safety
  1300 Connecticut Ave., NW
  Washington, D.C. 20036; telephone: 202-785-2525
Help for the Working Wounded, by Thomas F. Mancuso, M.D.; $1, 220 pp. Short and to the point about occupational

disease and its prevention, taken from Mancuso's Q & A column in *The Machinist*.
IAM Guide for Safety and Health Committees, $.25, 20 pp. Good on need for *independent* union-run committees in each shop.

LABOR OCCUPATIONAL HEALTH PROGRAM (LOHP)
   2521 Channing Way
   Berkeley, Ca. 94720; telephone: 415-642-5507
Monitor, monthly, $10/yr. institutions, $5/yr. individuals.
Working for Your Life: A Woman's Guide to Job Health Hazards (1977), $5 for individuals, $8 for institutions, 250 pp. First book of its kind.
Foundry Workers' Manual (Occupational Health and Safety), by Janet Bertinuson and Sidney Weinstein; $5, 100 pp.

LABOR SAFETY AND HEALTH INSTITUTE (LSHI)
   377 Park Ave. S. (27th St.), 3rd fl.
   New York, N.Y. 10016; telephone: 212-689-8959
An interesting one-man show by Frank Goldsmith. LSHI *Guides*, $4, 100 pp.

MASSACHUSETTS COALITION FOR OCCUPATIONAL SAFETY AND HEALTH (MassCOSH)
   120 Boylston St., Rm. 206
   Boston, Mass. 02116; telephone: 617-482-4283
Survival Kit, 6 times per year, $4. MassCOSH is a political action group, and is aided by:
   Urban Planning Aid
   2 Park Square
   Boston, Mass. 02116; telephone: 617-482-4283

UPA's publications are well written with excellent graphics—the first of their kind.
How to Look at Your Workplace, $.25, 27 pp.
How to Use OSHA, $.50, 100 pp.

Como Inspeccionar Su Centro de Trabajo, $.30, 40 pp. Best pamphlet I have seen on the subject for U.S. workers who speak Spanish. UPA also has bulletins on specific hazards.

OCCUPATIONAL HEALTH AND SAFETY INDUSTRIAL UNION DEPT., AFL-CIO
Shelly Samuels, Director
815 16th St., NW
Washington, D.C. 20006; telephone: 202-393-5581
SPOTLIGHT on Health and Safety (quarterly). Send name and address on a postcard and explain why you want it.

OIL, CHEMICAL, AND ATOMIC WORKERS (OCAW)
P.O. Box 2812
Denver, Colo. 80201; telephone: 303-893-0811
Lifelines (monthly), $5 per year to nonmembers of OCAW. Best union health and safety newsletter, written mostly by Sylvia Krekel.
Asbestos: Fighting a Killer, excellent 45-minute slide show with cassette tape, $125. Technically accurate and more of a treat to the eyes and ears than might be expected from the lugubrious nature of the material. Write Sylvia Krekel about possibilities of borrowing or renting.
Asbestos—a 3-foot silk-screened poster, $16.50 for union members and health specialists; $26.50 for corporations and other institutions.
Work Is Dangerous to Your Health, by Jeanne Stellman and Susan Daum (New York: Vintage, 1973), $2, 450 pp.

PHILADELPHIA AREA PROJECT ON OCCUPATIONAL SAFETY AND HEALTH (PhilaPOSH)
1321 Arch St., Rm. 607
Philadelphia, Pa. 19107; telephone: 215-568-5188
Safer Times, monthly, $5/yr., $12 for institutions; free to members.
Oil Refinery Health and Safety Hazards (1976), by Rick Engler, $3; $10 for institutions.

UNITED AUTO WORKERS (UAW)
  Social Security Dept.
  8000 E. Jefferson
  Detroit, Mich. 48214; telephone: 313-926-5321
  A Workers' Manual on Noise Control, Dan McLeod, ed.,
  $1.25.
  The Hazards of Lead and How to Control Them, $.75,
  21 pp.

U.S. GOVERNMENT
  The Occupational Safety and Health Administration
  (OSHA) is listed in the phone book under United States
  Government, Department of Labor; or the equivalent state
  agency. For technical pamphlets written for workers, the
  government can be a goldmine of technical information.
  Write:
  Publications
  US DOL/OSHA
  Room N-3423
  200 Constitution Ave., NW
  Washington, D.C. 20210; telephone: 202-523-8677

  NIOSH
  Division of Technical Services
  Public Dissemination
  4676 Columbia Parkway
  Cincinnati, Ohio

OTHER BOOKS
  Industrial Toxicology, by Alice Hamilton and Harriet
  Hardy, 3rd ed. (Publishing Sciences Group, Inc., 411 Mass.
  Ave., Acton, Mass. 01720, 1974), $26. Perhaps the best
  general physician's guide to the most common occupational
  diseases.
  Women's Work, Women's Health (Myths and Realities),
  especially chapter on stress and chapter called "Social Diseases . . . Social Cures." New York: Pantheon Books, $3.95,
  250 pp.

AUDIOVISUAL AIDS

Kepone, made for television show, "60 Minutes," about a pesticide-manufacturing atrocity in Virginia and the corporate attempts to evade responsibility and liability; $350 from CBS News, 524 W. 57th St., New York, N.Y. 10019. No mention of union involvement.

The Shop Accident, a corny movie about how to use an OSHA inspection to the worker's best advantage; 40 minutes, $220. Rent from AFL-CIO Film Library, 815 16th St., NW, Washington, D.C. 20006 for $12.50. Useful, despite klutzy presentation.

Are You Dying for a Job? 30-minute, 140-slide show with cassette tape. Slides are from actual industrial inspections, $125. Not bad—only thing of its kind. Write: Western Institute for Occupational/Environmental Science, 2520 Milvia, Berkeley, Ca. 94706; telephone: 415-845-6476.

INTERNATIONAL FEDERATION OF CHEMICAL AND GENERAL WORKERS (ICF)

This organization has sponsored a good deal of activity around occupational health issues, and has affiliates in dozens of countries around the world. Without strong union connections it is difficult to make contact with them, but worth the effort. Write:

Charles Levinson, Secretary-General
ICF
Secretariat: 58, rue Moillebeau
Geneva, Switzerland

Programs in Australia, Europe, and the Soviet Union

AUSTRALIA
Workers Health Center
Ben Bartlett
27, John Street
Lidcombe 2141, N.S.W.
Australia

DENMARK
  K.I.S.S.
  Lisbeth Bang
  Overgaden N. Vandet 51 B. St.
  1414 Kobenhaun K
  Denmark

FRANCE
  Collectif Intersyndical Securité
  CFDT-CGT-FEN
  Des Universités Paris 6 et Paris 7
  4, place Jussieu, Bâtiment H
  75230 Paris CEDEX 05
  France
For the last three years asbestos has been a subject of occupational health work. The Collectif has published a book on the subject entitled *Danger! Amiante* (40 francs), available from: Francois Maspero, 1 place Paul Painlevé, Paris 5e, France.

GREAT BRITAIN
  British Society for Social Responsibility in Science
  9 Poland St.
  London W1V 3DG
  Great Britain; telephone 01-437-2728
  Best publications:
  Hazard Bulletins, 20 pence or $.50; £2.50 per year for individuals; £5.00 for institutions; make checks payable to BSSRS Work Hazards Group.
  Noise, 25 pence, 34 pp.
  Oil (cutting oils, in U.S. terminology), 75 pence, 95 pp.
  Vibration (a workers' guide to the health hazards of vibration and their prevention), £1.00, 100 pp. Absolutely the best.
  All BSSRS publications are well laid out with excellent graphics. Write them for a list of health and safety projects in Great Britain.

ITALY
Medicina Popolare
Corso di Porta Roman, 55
20122 Milan
Italy

THE NETHERLANDS
Advisory Group on Health and Safety in the Netherlands
Peter Froenewegen
Van Swietanstraat 11
2334 EA Leiden
The Netherlands

SWEDEN
Joint Industrial Safety Council
Ingvar Soderstrom, Director
Sveagen 21, Box 3208, S-103 64
Stockholm
Sweden

USSR
Institute of Industrial Hygiene and Occupational Diseases
Professor N. F. Izmerov, Director
Budennogo Prospekt, 31
Moscow
USSR

Finally, I am available for help with any problems in the area of occupational health and safety. I can be reached at: 893 Rhode Island Street, San Francisco, Ca. 94107, telephone: 415-648-8274.

Daniel M. Berman

Index

Sullivan, Eva, 151
Supreme Court, U.S., 9, 60
Systems and Procedures Association of America, 178

Tabershaw, Irving R., 96–98
Taft-Hartley Act, 117–18
Taylor, George, 171
Teamsters union, 132–35, 173; Department of Safety and Health of, 133; Local 688 of, 96, 134; UFW and, 160
Texas, safety department expenditures in, 56
Tofany, Vincent L., 39–40, 43
Toxic Substances Control Act (1976), 69, 147
Trade Union of the City and County of Philadelphia, 25
Trasko, Victoria, 46
Traumatic neurosis, 65
Triangle Shirtwaist Company fire, 9
Tuberculosis, 29

U.S. Steel Corporation, 4, 10, 14; Clairton, Pa., coke plants of, 149; safety devices at, 77; suits against, 29–30; Voluntary Accident Relief Plan of, 15
Uniroyal plant, Eau Claire, Wis., 131
United Airlines, 143–46
United Auto Workers (UAW), 32, 125–30, 185; contractual right to strike over safety and, 119; Local 6 of, 31, 43, 98, 112
United Electrical, Radio, and Machine Workers (UE), 153–54, 173
United Farmworkers of America (UFW), 155–62
United Mine Workers (UMW), 136–42, 171; Anthracite Health and Welfare fund of, 138; black lung and, 31–32, 137–41; business ties of, 17, 137; NSC and, 139–39

United Papermakers and Paperworkers Local 800, 90
United Rubber Workers, 130–32
United Steelworkers of America (USWA), 32, 57, 147–53; CACOSH and, 112, 152
Upjohn Institute, 119
Uranium, ban on handling, 175
Urban Planning Aid, 35

Vela, Marcos A., 1–4
Vinyl chloride: exposure to, 35, 132; in Soviet Union, 193
Volkswagen, 183
Vos, J. G., 94

Wagner Act (1935), 159
Wallick, Frank, 126
Walsh-Healey law (1936), 58
Walter, Bob, 42
Washington, compensation in, 64
Weinstein, James, 26
Wells, Hawley A., 139
Wells, Kenneth F., 139
West Virginia, compensation in, 64
Western Federation of Miners, 19
Western Institute for Occupational/Environmental Sciences, 172
Williams, Harrison A., 36, 68, 72
Wirtz, Willard W., 56
Wise, Kent D., 1–2
Wodka, Steven, 123, 150
Wolkonsky, Peter, 107
Workers Action Movement, 128
Workingmen's Benevolent Association, 136
Worthington, Bill, 139
Wyoming, compensation funds in, 64

Yablonski, Jack, 140
Yocum, Elmer, 136
Young Communist League, 30

Z16.1, 41–43, 50, 80